Fundamentals of Applied Mechanics

Fundamentals of Applied Mechanics

C. THOMAS OLIVO
THOMAS P. OLIVO

DELMAR PUBLISHERS INC.

This volume is dedicated to Mr. and Mrs. Peter Olivo, the parents of the senior author, as a tribute to their drive, encouragement, and convictions about the tangible value to youth and adults of perseverance, industriousness, and the ever-present spirit of inquiry into the Why and How.

Delmar Staff:
    Administrative Editor: Jonathan Plant
    Production Editor: Carol Micheli

For information, address Delmar Publishers Inc.
2 Computer Drive West, Box 15015
Albany, New York 12212

COPYRIGHT © 1986
BY DELMAR PUBLISHERS INC.

All rights reserved. Certain portions of this work copyright © 1957, 1978, and 1984 under the title *Fundamentals of Applied Physics*. No part of this work covered by the copyright hereon may be reproduced or used in any form or by any means — graphic, electronic, or mechanical, including photocopying, recording, taping, or information storage and retrieval systems — without written permission of the publisher.

10   9   8   7   6   5   4   3   2   1

Printed in the United States of America
Published simultaneously in Canada
by Nelson Canada,
A division of International Thomson Limited

Library of Congress Cataloging in Publication Data

Olivo, C. Thomas.
    Fundamentals of applied mechanics.
    Includes index.
    1. Mechanics, Applied.   I. Olivo, Thomas P.
II. Title.
TA350.048   1986   620.1   85-20548
ISBN 0-8273-2625-4 (pbk.)

Contents

	BASIC PRINCIPLES UNITS	ASSIGNMENT UNITS
Preface	vii	
Introduction	ix	

SECTION 1 SCIENCE AND MATTER — BASIC PRINCIPLES

	BASIC PRINCIPLES UNITS	ASSIGNMENT UNITS
Unit 1 Science and the Scientific Method...............	1	6
Unit 2 The Science of Matter...........................	8	11
Unit 3 Basic Properties of Solids	14	16
Unit 4 Basic Properties of Liquids	18	21
Unit 5 Basic Properties of Gases........................	23	26
Achievement Review of Science and Matter	28	—

SECTION 2 SYSTEMS OF MEASUREMENT IN SCIENCE

	BASIC PRINCIPLES UNITS	ASSIGNMENT UNITS
Unit 6 Customary United States and British System of Measurement..........................	32	38
Unit 7 Metrication: SI Metric Units of Measure	41	56
Achievement Review of Systems of Measurement...	60	—

SECTION 3 MECHANICS, MACHINES, AND WAVE MOTION

	BASIC PRINCIPLES UNITS	ASSIGNMENT UNITS
Unit 8 Forces and Their Effects.........................	64	69
Unit 9 Balance and Equilibrium — Parallel and Angular Forces	72	79
Unit 10 Gravitation, Motion, and Mechanical Movements ...	82	91
Unit 11 Simple Machines: Levers.........................	95	100
Unit 12 Simple Machines: Inclined Plane and Wedge	103	107
Unit 13 Simple Machines: The Wheel and Axle	110	114
Unit 14 Simple Machines: The Screw Thread.............	117	123
Unit 15 Simple and Compound Gear Trains...............	126	131
Unit 16 Simple and Compound Machines.................	133	137
Unit 17 Mechanical Power Transmission, Friction, and Lubrication............................	139	149
Unit 18 Mechanics of Fluids at Rest and in Motion	153	158
Unit 19 Atmospheric Pressure: Principles and Applications..	163	170
Unit 20 Fluid Power: Principles and Applications	174	182

	BASIC PRINCIPLES UNITS	ASSIGNMENT UNITS
Unit 21 Wave Motion: Transfer of Energy	184	192
Achievement Review of Mechanics, Machines, and Wave Motion	195	—

APPENDIX

Table I	Standard Tables of Metric Units of Measure	206
Table II	Conversion Factors Using Customary and SI Metric Physical Constants	207
Table III	Symbols and Selected Derived Units of Physical Quantities	208
Table IV	Units of Mass and Weight and Conversion Factors	208
Table V	Comparative Densities of Gases Used in Industry	209
Table VI	Mass and Weight Densities of Common Substances	210
Table VII	Coefficients of Friction for Different Solids	210
Table VIII	Heat Values of Gases, Liquids, and Solids	211
Table IX	Physical Properties of Important Pure Metals	211
Table X	Tempering and Heat Colors	212
Glossary		213
Index		219

Preface

The scientist uses two powerful tools to control and use matter and energy. The first tool is that of reliable knowledge which has been verified and organized. The second tool relates to the acquiring and organizing of new knowledge and generalizations.

Scientific concepts revealed by investigation, exploration, and demonstration are utilized by skilled individuals who translate theory into fact. It remains for physicists, engineers, technologists, skilled craftspersons, and others, to transform the generalizations into inventions, products, processes, and systems which have widespread use.

This textbook is the end product of years of extensive enquiry, study, experimentation, and analysis of foundational understandings and the expertise in science required for career development and to meet the needs of each individual as a consumer. The contents cover concepts, principles, and applications in three areas of physical science: (1) the science of matter and the basic properties of solids, liquids, and gases; (2) measurement systems; and (3) mechanics, machines, and wave motion: transfer of energy.

In each of the *Sections* in the textbook there is a series of *Basic Principle Units*. The contents of these units were especially selected to provide instructional material for a concentrated study of principles. The scope of the units is such that the contents may be adapted to meet the special needs of students/trainees, allowing for progress according to varying individual abilities and desires. Within each unit there is a *Summary* followed by an *Assignment Unit* which provides a series of practical problems and applications of each basic principle. An *Achievement Review* at the end of each Section provides an enlarged, complementary bank of test items to comprehensively cover all the units.

Technical *Reference Tables* in the *Appendix* provide technical data, constant values, and formulas. *Tables of Conversion Factors* permit the easy conversion of a unit of measurement from either the *United States Customary* and *British System* or the *SI Metric System* to the other system. Whenever needed, it is practical by using simple arithmetical processes and a conversion factor to change a value in one system of measurement to its equivalent measurement value in another system.

The *Glossary* and *Index* (which are part of the *Appendix*) include major measurement quantities and terminology commonly used in business and industry.

A separate *Instructor's Guide* provides solutions to all *Practical Problems*. In addition, there is a *Comprehensive Review* covering the contents of all units in this book.

C. Thomas Olivo
Thomas P. Olivo

INTRODUCTION

Fundamentals of Applied Mechanics serves two basic purposes.
- The instructional units are written so as to present in a clear, functional way those principles and applications of physical science which deal with basic properties of matter in solid, liquid, and gaseous states; systems of scientific measurement; mechanics, machines, and wave motion: transfer of energy.
- The instructional units provide the student with fundamental understandings which are essential to the pursuit of advanced studies in related technologies.

■ ORGANIZATION AND SCOPE OF THE INSTRUCTIONAL MATERIAL

Fundamentals of Applied Mechanics contains three *Sections*. These are: (1) Science and Matter; (2) Systems of Measurement in Science, and (3) Mechanics, Machines, and Wave Motion: Transfer of Energy. Within each of these *Sections* there are a number of instructional Units. These are called *Basic Principle Units*.

■ BASIC PRINCIPLE UNITS

Each *Basic Principle Unit* is introduced with a brief statement giving the importance and relationship of each new topic or area of scientific investigation. This is followed by a planned presentation of background information which provides a base for formulating whatever scientific laws are considered in the unit. Examples are given to show how a law or physical condition may be applied in a practical way, thus giving meaning to each new principle.

The line drawings which appear throughout the units serve as teaching-learning devices either to place emphasis on an important point or to simplify a description.

Summary

Important new items are summarized at the end of each instructional unit for emphasis. The statements are brief and may also serve as a guide to the instructor in preparing each lesson.

■ ASSIGNMENT UNITS: PRACTICAL PROBLEMS

The *Assignment Unit* which complements each *Basic Principle Unit* contains a series of *Practical Problems*. For the most part, the problems are of the objective type and are arranged from the simple to the more complex. While computations are required to determine quantitative values, all problems have been kept comparatively simple. In this form, the problems emphasize and clarify specific principles and concepts and do not consume time in lengthy mathematical processes which may not reinforce the learning of a scientific principle. The units of measurement and the standards used throughout the text include the *Customary United States* and *British System* and the evolving *SI Metric System*. The use of these systems is consistent with current occupational practices.

In the explanations and examples provided in the text, measurements are converted between systems of measurement from Customary units to SI metric units. Generally, the exact values given in Table II in the Appendix are rounded off. In other cases, exact values are used where large quantities are involved and/or precise measurements are required.

■ SECTION ACHIEVEMENT REVIEWS AND COMPREHENSIVE ACHIEVEMENT REVIEW

An *Achievement Review* is provided to serve as both test and review material of the *Basic Principle Units* within a *Section*. The sequence of the problem material in each *Achievement Review* follows the same order as the *Basic Principle* and companion *Assignment Units* in that *Section*. The instructor may use the *Achievement Reviews* as pretests to measure student comprehension of certain basic concepts of science and their application in practical situations.

In addition to the *Section Achievement Reviews* in this textbook, there is a final *Comprehensive Achievement Review* in the *Instructor's Guide*. The *Comprehensive Review* provides a reservoir of supplemental test items.

■ RESOURCE TABLES, GLOSSARY, AND INDEX

The *Appendix* contains three items (1) a *Glossary* in which the major technical terms used in the text are further described; (2) a series of selected *Reference Tables* which make the text self-contained insofar as they provide technical data required in the solution of problems, and (3) an *Index* to assist in locating material within the text.

■ SUGGESTED APPLICATIONS FOR FUNDAMENTALS OF APPLIED MECHANICS

Each teaching-learning situation requires a different emphasis and presents a different need in the use of instructional materials. The following are ways in which the material in this text may be used effectively.

- As a science textbook for learners who must develop functional skills in science of matter; measurement; and mechanics, machines, and the transfer of energy and must learn how to apply these skills in practical problems that have meaning in an occupational career.
- As a basic science textbook of selected instructional units for organized class instruction for homogeneous groups or individualized self-paced instruction.
- As a basic, practical science textbook for work-experience related instruction, apprentice, specialized technical training, and in-plant occupationally-oriented courses where supplemental instruction is provided in other areas of science particular to an occupation.

- As a textbook or source book for adult programs and occupational extension classes where a concentrated, practical working knowledge of science up through mechanics may be directly related on-the-job to processes, tools, and materials.

■ INSTRUCTOR'S GUIDE AND ANSWER BOOK

The *Instructor's Guide* (which is a companion publication) contains the solutions to the objective test items in both the *Assignment Units* and the *Achievement Reviews*. This guide is intended to conserve valuable teaching time and to provide uniformity to the solutions of the problem material. A *Comprehensive Achievement Examination* is included in the Instructor's Guide as a final integrating experience covering all of the instructional units.

* * * * *

Fundamentals of Applied Mechanics incorporates a number of tested teaching-learning techniques for mastering basic scientific principles through simple, direct, and meaningful experiences. The instructional units develop an orderly working concept of the importance of science. They show how fundamental scientific laws, principles, and understandings may be applied with success to new technical developments that contribute so much to the attainment of higher standards of health, living, and welfare of all people.

Section 1
Science and Matter — Basic Principles

Unit 1 Science and the Scientific Method

Our technological age is being built by the peoples of all lands who seek to learn and apply fundamental scientific principles. It is this seeking out of truths and their applications which provides the scientist, the inventor, the technologist, and others with the basic knowledge which each needs to make possible experimentation, new developments, or production.

■ THE MEANING AND IMPORTANCE OF SCIENCE

The impact of science is evident everywhere. Disease has been controlled to such an extent that in the last fifty years the average life expectancy at birth has increased twenty years. Today, there is a greater abundance and variety of food; better transportation made possible by advances in design, materials, and fuels; faster communications; more comforts of life; better clothing; and better shelter. All of these improvements were made possible through new findings and gains in science.

Scientific *know-how* helps the artisan and technologist to manufacture materials and then shape or form them into parts which are useful in the home or on the farm; or into more complex or heavier mechanisms for use in business, industry, or research.

Science provides each person with simple truths about cause and effect. Each ingenious mechanism or new substance can be traced to a scientific beginning where some basic truths, proven through experimentation, were applied.

Science forms the foundation on which new substances and materials are produced and new products are built. The development of nylon, as shown in figure 1-1, is an example of the application of scientific facts.

Once the basic scientific fact was established that a fiber such as nylon was theoretically possible, continued laboratory experimentation produced the fiber; specialists and technicians developed the required plant and production facilities; and finally, nylon became a marketable product.

Within this framework, industry's principal contribution is to bring together three different areas of human effort — fundamental scientific knowledge, invention, and technology. Each of these areas is becoming increasingly more complex and specialized.

2 Science and Matter — Basic Principles

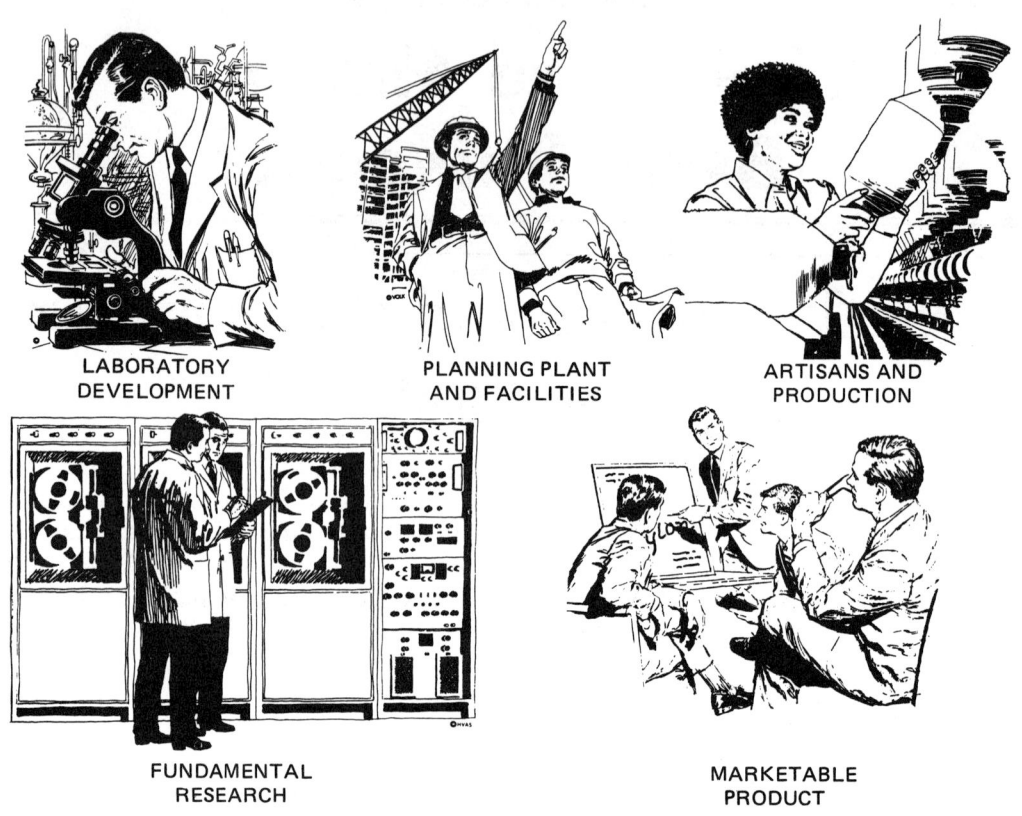

LABORATORY DEVELOPMENT

PLANNING PLANT AND FACILITIES

ARTISANS AND PRODUCTION

FUNDAMENTAL RESEARCH

MARKETABLE PRODUCT

Fig. 1-1

For example, consider a single product such as a chemical weed destroyer. To produce this chemical, scientists, inventors, and technologists from many different branches of science, such as biochemistry, plant physiology, pathology, soil chemistry, and others, must work together. Such combinations of people are common and are needed to create, test, experiment, produce, and market the goods of the world. Continuous production is another example of how science is combined with experimentation and manufacturing capability.

■ **BRANCHES OF SCIENCE**

Communication, transportation, manufacturing — physical, chemical, and biological developments — all depend on science and the application of the scientific method.

Science is *organized knowledge* derived from the use of a systematic approach to a problem and the application of the scientific method. Physical science requires planning, experimentation, observation, analysis, and problem solving under controlled conditions. Scientific principles, laws, and philosophy are based on broad generalizations.

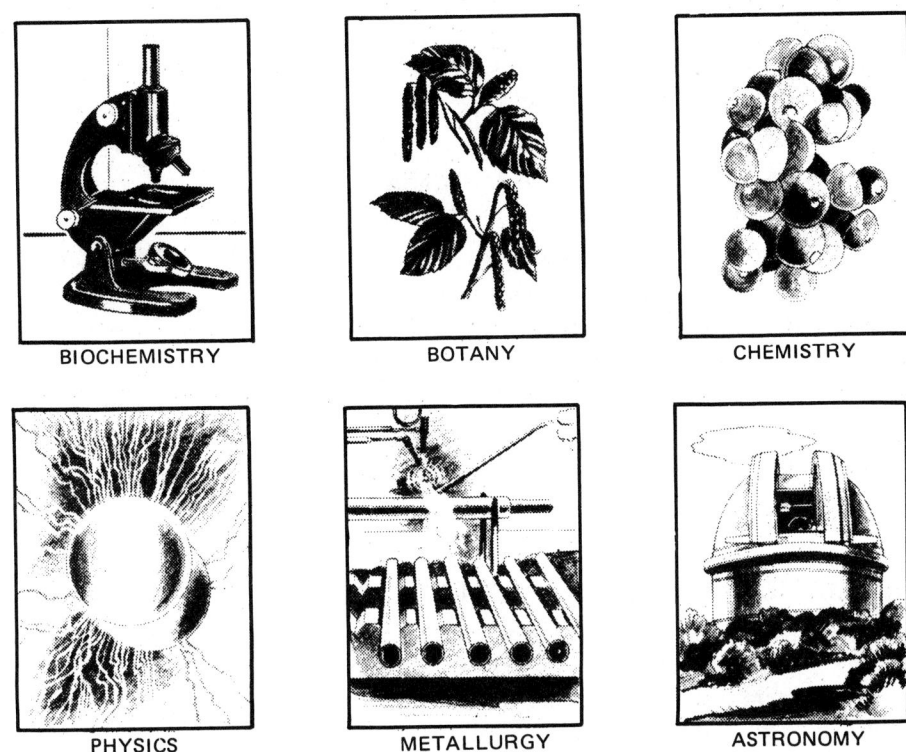

Fig. 1-2

Every generalization is subject to modification. Prior deductions must be continuously assessed in the light of new evidence. Thus, science is a living and evolving body of information.

Scientific knowledge is classified into specific groupings, figure 1-2. *Biology* or the *biological sciences* deal with life and living things. *Botany* is a branch of biology dealing with plant life. *Zoology* is a scientific study of animal life.

Scientific knowledge may also be grouped around nonliving things like the sun, moon, stars, and planets. Information about these celestial bodies belongs in the area of study called *astronomy*.

Chemistry deals with the changes that occur in the composition of matter. *Physics* is a study of matter. (Matter is anything that takes up space and has mass and energy. Energy is the ability to produce work.) *Metallurgy*, another branch of the physical sciences, deals with the study of the properties of metals, their grain structure, and the effects of adding metals to one another.

The student should be aware of the dependence on science of engineering and technological developments, the national economy and security, standards of living, and life itself. In particular, the science of physics is considered to be the basic science whose principles and concepts are drawn upon by many other fields of science.

■ THE SCIENTIFIC METHOD

The *scientific method* is a systematic procedure for discovering basic truths. There are five simple steps to the scientific method.

- Recognize the problem to be solved and the essential parts of the problem.
- State a set of hypotheses to solve the problem. *Hypotheses* are possible explanations or scientific conditions which are the cause of the observed results.
- Collect facts to test each possible explanation or hypothesis. These facts may be obtained by experimentation or any other method.
- Analyze and interpret the data (facts) to determine the correctness of each possible explanation (hypothesis).
- Test conclusions in enough new situations to be sure they are correct.

The scientific laws of physics, chemistry, biology, and the other sciences, which are widely used and accepted today, are the result of experimentation and tests based on the scientific method.

■ SYSTEMATIC APPROACH TO PROBLEM SOLVING

Scientific investigation requires that specific procedures be followed in an orderly manner. This practice enables the investigator to analyze a problem, to gain a perspective of the magnitude and conditions of the problem, to prevent errors, and to save time. Scientific problem solving also requires the systematic processing of data to a specified degree of accuracy. The solutions are stated in a significant number of figures. Alternate computational methods are used to recheck the solutions.

The following steps are recommended for the solution of physical science problems. In addition, these steps can be applied effectively to the solution of other practical occupational problems.

1. Read the statement of the problem carefully.
2. Determine the nature and degree of accuracy required in the solution. This step is necessary to avoid extra numerals in a computation.
3. Draw an appropriate diagram, listing the given data.
4. Determine and state the physical principles which appear to be relevant to the solution of the problem.
5. Determine if all of the required data are available. Identify the sources or methods that may be used to obtain missing data.
6. When using formulas, determine if the mathematical processes are to be carried out by first using symbols and, later, substituting numerical values, or if number values are to be used immediately.
7. Check the units for each quantity to be used in the problem to insure that all units are in the same system of measurement.

8. Substitute the data obtained from the physical principles in the problem.
9. Compute the required numerical value. Record the result so that only the significant numerals (figures) are retained.
10. Determine the units in which the results are to be expressed and label the results accordingly.
11. Examine the result to see if it is a reasonable solution. Determine if an alternate method of checking may be used.
12. Recheck all values, formulas, and computations by reworking the problem using an alternate method.

This modern study of science begins with the branch of science called physics. Each unit that follows states the basic principles or laws of physics which are derived by applying the scientific method. In the Student's Guide, experiments are suggested to give meaning to the study of specific scientific principles. When results are obtained from performing one or more of the steps given in the procedure for each experiment, these results should be put in table form. As these pieces of data become complete, they are interpreted to identify and explain the scientific principle being demonstrated.

This opportunity to apply scientific principles, using the scientific method, should make the study of science both practical and interesting.

SUMMARY

- The seeking out of scientific truths is the foundation on which the scientist, inventor, and technologist control the physical world and shape it to meet human needs.
- Science is the gathering, testing, and arranging of organized knowledge.
- Scientific knowledge is grouped around a specific science and may deal with living things or nonliving matter.
- The fundamental findings of research in various branches of science may be combined to test, improve, finance, produce, and market the goods of the world.
- Industry's greatest contribution to mankind has resulted from bringing together three different areas of human effort: science, invention, and technology.
- Physics deals with a study of matter and energy and is the science upon which other fields of science are based.
- The scientific method, together with other organized procedures, is essential to any study, research, demonstration, and application of physical science principles and concepts.

Science and Matter – Basic Principles

ASSIGNMENT UNIT 1 SCIENCE AND THE SCIENTIFIC METHOD

■ PRACTICAL PROBLEMS ON THE MEANING AND IMPORTANCE OF SCIENCE

Science in the Nuclear Age

For statements 1 to 8, determine which are true (T) and which are false (F).

1. New production, developments, and experimentation are made possible largely through the application of fundamental scientific principles.
2. Gains in science have not affected the standard of living or the life span of an individual.
3. Scientific laws and principles help the inventor, technologist, and scientist to produce and shape materials into useful products.
4. Industry and business contribute to society and scientific progress by welding the following areas into a productive team: fundamental scientific knowledge, invention, and technology.
5. There is no dependence of one branch of science on another.
6. Scientific laws and principles are applied daily to living things and nonliving matter.
7. The scientific method has no influence on the development of scientific principles.
8. The scientist, inventor, or technologist seeks out scientific truths as the foundation for controlling the physical world and shaping it to meet human needs.

Complete statements 9 to 15 by adding a word of phrase.

9. Scientific knowledge is gained through _____ and _____.
10. The _____ sciences deal with life and living things.
11. _____ is a study of animal life.
12. _____ is that branch of science dealing with the solar system.
13. Physics is a study of _____ and _____.
14. Wherever metals are used, a physical science such as _____ deals with their properties and fabrication.
15. _____ is a study of the composition of matter.

Select the correct word to complete statements 16 to 19.

16. (Botany) (Zoology) is a study of plant life.
17. The study of animal life is called (zoology) (bacteriology) (astronomy).
18. Chemistry is a study of the (energy) (life) (composition) of matter.
19. Biology is the science of (matter) (life) (metals).

20. Identify three additional branches of science that are not mentioned in this unit.
    a. Name each branch of science.
    b. Describe briefly what each branch covers.
21. Study one new industrial product.
    a. Name the product.
    b. Identify the kind of scientific knowledge which any two branches of science contribute to the development of the product.

The Scientific Method
1. List the five basic steps in the scientific method. Be brief.

Add a word or phrase to complete statements 2, 3, and 4.
2. Scientific facts may be obtained by _____.
3. Facts must be _____ and _____ in the scientific method.
4. A scientific set of hypotheses means _____.

Select the correct word or phrase to complete statements 5 and 6.
5. A conclusion tested in (a single instance) (a sufficient number of new situations) is an example of a good scientific technique.
6. The (facts) (estimates), when analyzed, determine the correctness of a basic hypothesis.
7. List three values to be derived from a study of physics, chemistry, and industrial materials.

■ SYSTEMATIC APPROACH TO PROBLEM SOLVING

Identify a simple physical science problem.
1. Review each of the twelve steps in the systematic approach to problem solving.
2. Analyze the problem and state how each step may be used to solve the problem.

Unit 2 The Science of Matter

Physics is the branch of science that deals with solids (such as steel and plastics), liquids (oils and water), and gases (hydrogen and the other gases in air). The physicist calls these various types of material matter. Thus, *matter* is anything that occupies space and has weight.

Steel, oil, and air are three different states or forms of matter. The steel is a solid; the oil is a liquid; and the air is a gas. Each form or state of matter is identified by certain characteristics which are called *properties*.

This unit deals with the structure of matter and, in a general way, with the common properties of matter.

With this kind of understanding as a starting point, later units cover specific principles that apply to solids, to liquids, and to gases.

■ STRUCTURE OF MATTER

All matter is made of small particles called *molecules*. Each molecule is actually the smallest particle of a material which retains the properties of the original material. For example, if a grain of table salt is divided in two, then each subsequent grain is divided again, and the process is continued as finely as possible, the smallest particle having all the properties of salt is a molecule of salt. This molecule is almost one millionth of an inch in diameter. The molecule is so small that before it can be seen in a microscope, it must be enlarged about 100 times. As small as the molecule may seem to be, nuclear physicists work with even smaller particles.

Properties Depend upon Molecular Arrangement

The molecules in a given material are all alike. However, different materials have different molecules. The characteristics and properties of different materials depend upon the nature and arrangement of molecules. For example, the differences in the weights, colors, and hardnesses of two metals all depend upon the arrangement and structure of their molecules.

When the form of a material is changed, the change is referred to as a *physical change*. This means that although a material is pressed, cut, or in some way has its shape modified, no new material is produced because the molecular structure remains the same.

■ PHYSICAL PROPERTIES OF MATTER

The properties of matter are described and classified by terms that have very special meanings. Some of the common terms are: mass, weight, volume, density, porosity, cohesion, adhesion, impenetrability, and inertia. Later, in the more advanced units dealing with matter, additional terms will be introduced and described. The terms

The Science of Matter 9

listed in this unit provide a working understanding of matter. The mathematics used in making measurements of the properties of matter will be covered in another unit.

Mass and Weight

The *weight* of a body is the pull of the earth on the body, figure 2-1. The weight of a body is a variable quantity which depends in part on the distance of the body from the earth's center. Since the earth is not a perfect sphere, a man weighing 160 pounds (72.6 kilograms) in New York will weigh 161 pounds (73.1 kilograms) at the North Pole. At the pole, he is about 12 miles (19.3 kilometers) closer to the center of the earth.

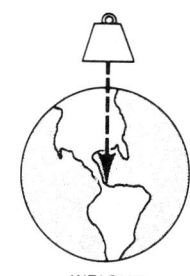

EARTH'S PULL

WEIGHT

Fig. 2-1

Assume that a stone weighing six pounds on the earth is taken to the surface of the moon. The moon attracts this stone with a force of about one pound. Thus, the weight of the stone on the moon is only one pound. Although it is easier to keep the stone from falling, the force that one must use to throw the stone fast is the same as it is on the earth. If someone is hit by the thrown stone, it hits just as hard as it does on the earth at the same speed.

The behavior of the thrown stone is said to be due to its *mass*. The mass of a body is the amount of material in it, as demonstrated by its opposition to a change in speed. An object always has mass; this mass is usually a constant quantity, independent of the object's position. Mass is used in scientific calculations of forces, energy, and speed changes.

- Mass is a measure of the amount of material in a body.
- Mass is not affected by the distance of the body from the earth's center.

Volume

Volume is the space that a body occupies, figure 2-2. A cube of metal one foot on a side has a volume of 1728 cubic inches. This means that the total space the piece of metal occupies is a specific number of measurement units which remains the same regardless of where the metal is placed.

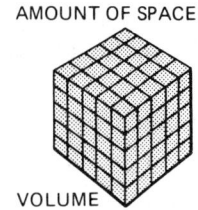

AMOUNT OF SPACE

VOLUME

Fig. 2-2

Density

Materials are often said to be light or heavy. Actually, these words mean that in comparing the weights of two equal volumes, one is heavy or light in proportion to its size, figure 2-3. *Density* is the mass for a unit volume. Tables of density are available which indicate the weight for a specific volume for many types of materials.

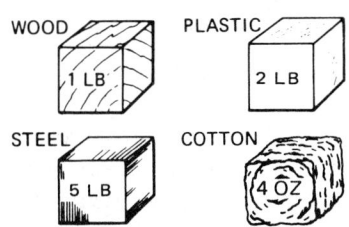

Fig. 2-3 Comparison of densities

10 Science and Matter — Basic Principles

For example, lead weighs 700 pounds (317.8 kilograms) per cubic foot and cork weighs 15 pounds (6.8 kilograms) per cubic foot. There is a wide variation between the densities of these two materials for the same volume.

Porosity

Porosity describes the structure or arrangement of molecules in a material. The greater the spaces between the molecules of the material, the more *porous* it is said to be, figure 2-4.

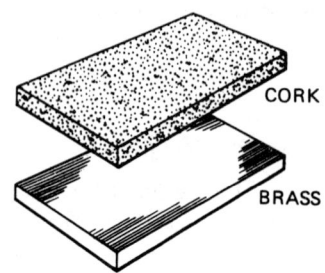

Fig. 2-4 Porosity

Cohesion

The degree to which the molecules in one body attract the molecules in another body is described as *molecular attraction*. The molecules in solids do not separate but have an attraction for each other. This tendency of two pieces of the same material to stick together is called *cohesion*, figure 2-5. Without cohesion, solids and liquids could not exist. It is the property of cohesion between molecules that holds solids and liquids together in a recognizable form.

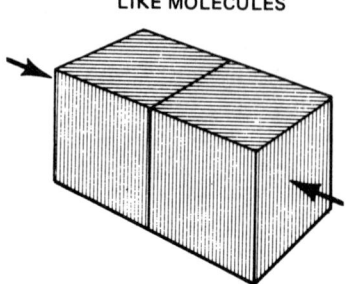

Fig. 2-5 Cohesion

Adhesion

Some molecules have a tendency to stick to molecules of a different material, figure 2-6. This property is called *adhesion* and is the attractive force between unlike molecules.

Impenetrability

All matter has another property, that of impenetrability. *Impenetrability* means that two bodies cannot occupy the same space at the same time. It is apparent that two blocks of steel cannot occupy the same place at the same time.

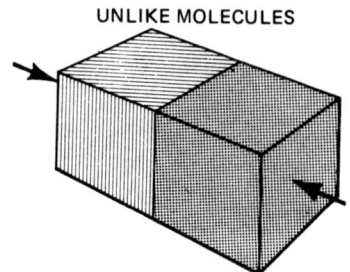

Fig. 2-6 Adhesion

Inertia

Force is required to start, stop, or change the direction of motion of matter. Objects such as tables or chairs remain at rest until a pushing or pulling force is applied. On the other hand, a car or a ball in motion tends to keep moving until a force changes the speed or direction of motion. The reluctance these objects show to a change is called *inertia*. The greater an object's mass, the greater is its inertia.

These, then, are the properties of matter: mass, weight, volume, density, porosity, cohesion, adhesion, impenetrability, and inertia. The more specific properties of solids, liquids, and gases are described and applied in later units.

SUMMARY

- The scientist and technician consider matter as anything that occupies space, is perceptible to the senses, and has weight and mass.
- Matter appears in any one of three forms: solid, liquid, or gas.
- The molecule is the smallest particle of a substance that retains the properties of the original substance.
- Mass is the amount of material in a body. Mass is constant and is not affected by the distance of the body from the earth.
- Weight is the pull of the earth on a body and varies with distance from the center of the earth. Depending on an object's location, weight may also be due to the pull of any celestial body.
- Volume is the space or room that a body occupies.
- Density is the mass per unit volume.
- Porosity refers to the spacing or arrangement of molecules in a material.
- Cohesion is the tendency of molecules of the same material to attract or cohere to each other.
- Adhesion is the ability of unlike molecules to attract or adhere to each other.
- Impenetrability is that property of all materials to occupy space.
- Inertia is the reluctance of a body to change its speed, direction of motion, or position of rest.

ASSIGNMENT UNIT 2 THE SCIENCE OF MATTER

■ PRACTICAL PROBLEMS ON THE STRUCTURE AND PHYSICAL PROPERTIES OF MATTER

The Structure of Matter

For statements 1 to 5, determine which are true (T) and which are false (F).

1. Matter may exist in any one of three states: solid, liquid, or gas.
2. All of the molecules in a given material are not alike.
3. The arrangement and structure of molecules determine the properties of different materials.

12   Science and Matter — Basic Principles

4. The physical working of a material produces a physical and not a chemical change.
5. The smallest particles of matter that scientists experiment with are molecules.

Physical Properties of Matter
1. Define briefly what is meant by mass.
2. Which of the two properties, mass or weight, may vary? Give reasons.

Select the correct word or phrase to complete statements 3 to 6.
3. The space that a body occupies refers to its (volume) (density).
4. The arrangement of the molecules in a body determines its (inertia) (adhesion) (porosity).
5. The ability of the molecules of the same material to attract each other is (adhesion) (cohesion) (impenetrability).
6. (Density) (Adhesion) refers to the ability of unlike molecules to attract each other.

Indicate the correct physical term to complete statements 7 to 10.
7. _____ is the amount of material in a body and is not affected by distance from the earth's center.
8. The resistance of a body to change in motion is known as _____.
9. Matter has the property of _____, which means that two bodies cannot occupy the same space at the same time.
10. A material that is less porous than cork is _____.
11. On a separate sheet, match each term in Column I with the correct physical property described in Column II.

Column I	Column II
1. Impenetrability	a. A comparison of the weight of equal volumes of matter.
2. Cohesion	b. The amount of matter in a body.
3. Adhesion	c. The pull of the earth on a body.
4. Mass	d. The property of matter to occupy space.
5. Density	e. The ability of unlike molecules to attract each other.
	f. The ability of like molecules to attract each other.

12. Select three common materials and compare some of their physical properties.
    a. Prepare a table like the one illustrated.
    b. List the three materials.
    c. Identify and record the color of each material.

The Science of Matter 13

d. Compare the porosity of the three materials and record the results using a 1, 2, 3 scale.
e. Compare and record the density of the three materials, using the same scale.
f. Estimate and record how the three materials compare for cohesive properties.

Materials	Comparison of Physical Properties			
	Color	*Porosity	*Density	*Cohesion
A				
B				
C				

*Use a 1, 2, 3 rating scale: (1) to indicate the greatest or most; (2) next; (3) the least.

Unit 3 Basic Properties of Solids

Matter exists in three states: solid, liquid, or gas. In each state, matter has specific properties. These properties are used to identify matter and determine its use. The properties to be studied in this unit are those of solid matter.

A solid consists of molecules arranged so that they cohere; that is, they attract each other so strongly that they stay fairly close together. This cohesion gives the solid definite size and shape and affects such properties as hardness, elasticity, and machinability. These and other properties of solids will be analyzed and defined. Thus, as a solid is considered from this point on, it will be understood in terms of its properties.

■ HARDNESS, TOUGHNESS, MALLEABILITY

Hardness is the ability of a material to resist forces that tend to push the molecules apart. Hard materials resist being scratched, worn away, penetrated, or indented.

Examples: Industrial diamonds, which are harder than rocks, are used in well drilling.

Abrasive wheels are used to cut away metals, plastics, and nonmetallic materials. Cutting of this type is possible because the individual abrasive particles are harder than the material to be cut.

Toughness is the property of a material that enables it to withstand a permanent change.

Example: Laminated wood is used on building trusses because of its toughness. This property helps prevent collapse when the structural member is loaded.

Malleability is the property of a material which deals with its ability to withstand mechanical processing such as hammering and rolling.

Example: When an iron casting is chilled in cooling, it is brittle. This same casting can be heat treated until it is malleable enough to be hammered into another shape without breaking.

HARDNESS

TOUGHNESS

MALLEABILITY

Fig. 3-1

■ DUCTILITY, ELASTICITY, TENACITY

Ductility refers to the ability of a material to be drawn into shape without losing other mechanical properties, figure 3-2.

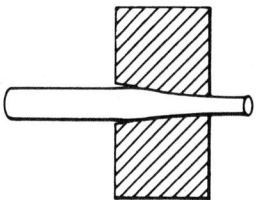

Fig. 3-2 Ductility

Example: In wire drawing, a metal rod is drawn through progressively smaller openings until the desired size is reached. While it is being drawn, however, the wire retains its physical properties.

Elasticity is the tendency of a material to return to its original form when a deforming force is removed. Once a pattern of molecules is formed in a solid, these molecules seek to return to this pattern after the force tending to push them out of the pattern is removed.

Example: A structural steel girder must be elastic so that it will return to its original shape after a load is removed.

Tenacity is the cohesive ability of a material to resist forces which tend to pull the material apart.

Example: The tenacity of the fine strands of wire in a large bridge cable enables them to withstand the tremendous loads encountered in supporting a bridge.

■ MACHINABILITY, FUSIBILITY, CONDUCTIVITY

Machinability refers to the ease with which a material can be shaped by cutting tools.

Example: The machinability of aluminum is greater than that of tool steel. Machining operations such as drilling and turning can be performed more easily and quickly on aluminum than on tool steel.

Fusibility is the ability of a solid to change from the solid to a liquid state when heated. This term also applies to the ease with which a solid may be united or fused with another material.

Example: The fusibility of welding rod with certain metals makes the joining of parts by welding possible.

Conductivity refers either to heat conductivity or electrical conductivity. *Heat conductivity* is the ability of a material to permit the flow of heat through it. *Electrical conductivity* is the ability of a material to permit the flow of electricity.

Examples: Copper is a good conductor of heat and electricity.

Wood is a poor conductor of both heat and electricity and is used as an insulation against them.

16 Science and Matter — Basic Principles

```
_____ SUMMARY _____
```
- Every solid consists of molecules held together by cohesion.
- The cohesive force gives each solid its shape and size.
- The abilities of the molecules in a solid to resist physical and electrical changes are described as the properties of the solid.

Property	Characteristics of Solid Due to Property	
Hardness	Resist scratch and wear	
Toughness	Withstand permanent change	
Malleability	Take mechanical processing	
Elasticity	Return to original form	
Ductility	Be pulled to shape	
Machinability	Be shaped by cutting tools	
Tenacity	Resist forces pulling material apart	
Fusibility	Change shape and unite	
Conductivity	Heat	Permit flow of heat
	Electricity	Permit flow of electricity

```
_____ ASSIGNMENT UNIT 3 BASIC PROPERTIES OF SOLIDS _____
```

■ PRACTICAL PROBLEMS ON PROPERTIES OF SOLIDS

Hardness, Toughness, Malleability

Select the correct word or phrase to complete statements 1 to 4.
1. A steel saw is used to cut wood because the saw is (harder) (softer) than wood.
2. (Machinability) (Toughness) is the ability to withstand permanent change.
3. A brittle casting is (more) (less) malleable than a soft lead plate.
4. Heat treating a hardened steel part or chilled cast iron casting makes it (brittle) (tougher).
5. List three metals in the order of their hardness.

Ductility, Elasticity, Tenacity

For statements 1 to 4, determine which are true (T) and which are false (F).
1. Ductile materials may be drawn into shape easily.
2. Elasticity is the ability to withstand hammering and rolling.

3. Elasticity is the ability to resist forces that tend to push molecules apart.
4. A ductile material retains its physical properties even when drawn to a smaller size.
5. Describe (a) elasticity (b) tenacity.

Machinability, Fusibility, Conductivity

1. On a separate sheet, match the properties in Column I with the material in Column II that has the required properties.

    Column I   Column II
    1. Machinable   a. Abrasive grinding wheel
    2. Fusible   b. Bronze rod
    3. Conductor   c. Silver wire
                           d. Solder

2. Describe: (a) conductivity (b) fusibility
3. List three materials that are good heat conductors.

Unit 4 Basic Properties of Liquids

A different set of properties is used to describe the characteristics of matter in the liquid form. Terms such as surface tension, solvent, buoyancy, specific gravity, and capillary action are associated with liquids. This unit identifies and illustrates those properties which the student must understand to be able to work with liquids.

■ COHESION, ADHESION, SURFACE TENSION

Cohesion

A liquid has the properties of cohesion and adhesion, but to a lesser degree than solids. The ability of like molecules in a liquid to *cohere* is the property that holds the liquid particles together.

Adhesion

Adhesion is the tendency of different molecules to stick together. For example, water "wets" some metals because the unlike molecules of the water and the metal stick together. There is less wetting action between water and oil because the molecules of these liquids are less adhesive. If water did not have the property of adhesion, it would not "wet" a surface.

Surface Tension

The molecules on the surface of a liquid are pulled downward by cohesion. Thus, the surface tends to contract. *Surface tension* refers to the tendency of the surface of a liquid to shrink and become as small as possible. Surface tension also makes the surface of a liquid seem tougher than the interior.

Capillary Action

The three properties of cohesion, adhesion, and surface tension in combination produce capillary action. *Capillary action* generally describes how a liquid behaves when its surface is in contact with a solid.

Liquids with Greater Adhesion than Cohesion Properties

Capillary action can be explained with the aid of a simple demonstration. If a glass tube with a small bore is placed in a container of water, the water will rise in the tube, figure 4-1. This action results from the fact that the cohesion and adhesion of a liquid are seldom equal and the surface tension

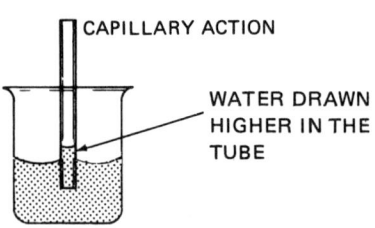

Fig. 4-1 Liquid with greater adhesion than cohesion

tends to shrink the surface. In this case, the adhesion of water to the glass is greater than the cohesion between the molecules of water. Thus, the water wets the glass surface and moves up the walls of the tube.

As the water crawls up the walls of the tube, the surface of the water becomes curved. Surface tension (which is trying to contract the surface and keep it as flat as possible) pulls the water up under it.

The water in the tube rises until the weight of the water under the surface film is equal to the combined adhesive and surface tension forces. The smaller the surface area of the water in the tube, the greater are the adhesive and surface tension forces and the higher the water will be lifted before its weight equals the forces.

Liquids with Greater Cohesion than Adhesion Properties

Not all liquids show the same capillary action. For example, the cohesion of mercury is greater than its adhesion. As a result, instead of crawling up the walls of a tube or container, mercury seems to crawl downward, figure 4-2. Since the surface tension and cohesion of the mercury molecules are greater than the adhesive force, the molecules tend to stick to one another rather than to a different substance. The result is that the mercury in the tube drops to a lower level than the mercury in the container. The surface of the mercury both in the tube and the container is mounded (convex).

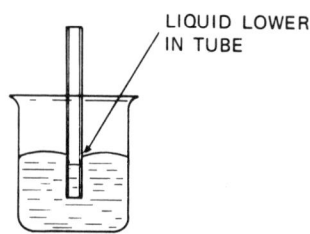

Fig. 4-2 Liquid with greater cohesion than adhesion

■ LIQUIDS AND EMULSIONS

The properties of adhesion and cohesion are important in industrial applications. High production is possible because *coolants* may be used to reduce the heat produced in a machining operation. The coolant wets the work and clings to it to remove heat.

Water is a good coolant because it is cheap and has the ability to remove heat effectively. However, water tends to produce rust. A mixture called an *emulsion* is produced by adding oil to water. This oil contains a special soap that allows the oil to mix with the water. This type of oil is said to be *miscible*. The milky-appearing water and oil emulsion combines the cooling properties of water with the rust prevention properties of oil to make an ideal coolant.

In draining the crankcase of an automobile, the sludge that is removed is an emulsion of oil, water, and dirt particles.

■ VISCOSITY

Viscosity refers to the resistance of a fluid to the motion of bodies through the fluid or to currents in the fluid. The viscosity of fluids is important because the friction that must be overcome due to viscous drag requires the expenditure of a great amount of energy. In the design of a system, viscosity must be considered so that materials, parts, and mechanisms can be selected and shaped to move in or through a fluid with a

20 Science and Matter — Basic Principles

minimum of resistance. When the design involves liquids, the pressures exerted on the liquid and the operating temperature are other design factors. For example, automotive and aircraft parts require lubricants of different viscosities and properties. There are variations of viscosity even with the same liquid. At a constant temperature, a No. 10 grade oil offers less resistance to flow than a heavier oil. The important consideration in selecting a lubricant is that the viscosity of the oil must insure that contacting surfaces are always separated by an oil film.

■ BUOYANCY AND SPECIFIC GRAVITY

Buoyancy

Liquids exert a force against materials which are placed in them. Because of the impenetrability of materials, an object placed in water pushes the water aside and begins to sink. When the weight of the water displaced (pushed aside) by the object equals the weight of the object, the object floats. When the weight of the object is greater than the weight of the water displaced, the object sinks.

The tendency of the liquid to produce an upward force on an object in the liquid is known as *buoyancy*. The amount of this buoyant force equals the weight of the liquid displaced.

Specific Gravity

Water is taken as the standard of comparison when the buoyancy of an object is to be determined. A cube of water twelve inches (30.48 cm) on a side weighs 62.4 pounds (28.3 kg). An object with a weight per volume equal to that of water will float. This relationship is known as the *specific gravity* of the material. For water, the specific gravity is 1.0. An object with a specific gravity greater than 1.0 will sink; one whose specific gravity is less than 1.0 will float, figure 4-3.

Specific gravity, then, is stated as a number (a ratio) which indicates the relationship between the weight of a given volume of material and the weight of an equal volume of water. For example, if a cubic foot of one liquid weighs 93.6 pounds and a cubic foot of water weighs 62.4 pounds, the specific gravity of the liquid is the ratio 93.6/62.4 or 1.5. This value means that the liquid is 1.5 or 1 1/2 times heavier than water.

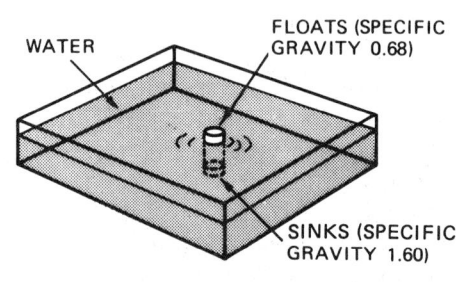

Fig. 4-3

SUMMARY

- Molecules in liquids have cohesive properties.
- The property of adhesion causes many liquids to wet the surfaces of solids.

Basic Properties of Liquids

- Surface tension produces a surface that appears tougher than the interior and is shrunken in size.
- Capillary action depends on cohesion, adhesion, and surface tension.
- Capillary action is the result of the interaction of these three properties. This action continues until the column height of the liquid is equal to the combined adhesive and surface tension forces.
- Practical applications of cohesive properties are found in emulsions which are widely used in manufacturing, agriculture, business, and the home.
- Viscosity is that property of the molecules of a fluid which resists the motion of bodies or currents through or in the fluid.
- Buoyancy is the ability of a liquid to exert force and support an object.
- Water, with a specific gravity of 1.0, is used as a standard of comparison when stating the specific gravity of a material.
- The number used to indicate the specific gravity of a material expresses the ratio of the weight of a volume of the material to the weight of the same volume of water.

ASSIGNMENT UNIT 4 BASIC PROPERTIES OF LIQUIDS

■ PRACTICAL PROBLEMS WITH PROPERTIES OF LIQUIDS

Cohesion, Adhesion, Surface Tension

1. Describe briefly what is meant by (a) adhesion, (b) cohesion, and (c) surface tension.
2. Name four common liquids used in industry or business.
3. For each of the four liquids listed in problem 2, state whether the property of cohesion or adhesion is greater.

Capillary Action

1. Using a simple cross-sectional sketch, show the capillary action of a liquid that has greater properties of cohesion than adhesion. Label the sketch.

Solutions, Emulsions and Viscosity

Complete statements 1 to 4 by adding the correct term for each.

1. Heat may be carried away in an industrial process by a _____.
2. A mixture of water and oil is referred to as a (an) _____.
3. _____ refers to the ability of the molecules of a liquid to resist a change in shape.

22   Science and Matter – Basic Principles

4. The heavier the grade of an oil, the greater is the _____ offered to flow.
5. Describe briefly what viscosity means.
6. Give one example each of an important application of viscosity in an automobile and in an airplane.

Buoyancy

1. Indicate which of the following materials will float in gasoline.
   - a. Cork
   - b. Bronze
   - c. Aluminum
   - d. Lead
   - e. Copper
   - f. Magnesium
   - g. Polyethylene
   - h. Slate
   - i. Wood
2. Name three industrial materials that sink in water.
3. List three nonmetallic materials that float in oil.

Specific Gravity

For statements 1 to 5, determine which are true (T) and which are false (F).

1. Mercury is the standard on which the specific gravity of materials is based.
2. Specific gravity is the relationship between the weight per volume of the material and water.
3. A material with a smaller specific gravity than water floats.
4. A material with a specific gravity of 0.75 sinks in water.
5. The specific gravity of water is 1.0.
6. List five materials that have a specific gravity greater than water. A listing of materials can be found in a handbook table on Specific Densities of Materials. The specific gravity can also be determined by laboratory tests.

Unit 5 Basic Properties of Gases

Gases have many properties which are different from those of liquids or solids. The student should have an understanding of the common properties of gases such as weight, fluidity, pressure, and density.

■ MOLECULAR MOTION IN GASES

Fluidity of Gases

Gases have the ability to expand indefinitely or to be compressed into a smaller volume. When heated, a gas expands and becomes less dense. This action is explained by the fact that the molecules of gases move farther apart when heated than they do when cool. Heated gases are lighter than cooler gases and so they tend to rise. The *fluidity* of air (its ability to flow in a manner similar to a liquid) is brought about by changes in pressure and variations in temperature. This fluidity causes the common storm.

The molecules of a gas may move freely in any direction. Single molecules move at high speeds. Although the movement of one molecule produces little effect, large numbers of molecules acting on a container may produce a pressure.

This principle can be explained by noting the increase in pressure when a closed container of air is heated. As the container is heated, the speed and distance between the molecules increase and a pressure builds up. As the source of heat is removed, the molecules slow down and move closer together again. If any air has escaped from the container, the air pressure outside the container exerts a force against the container and causes it to collapse. When the pressure inside and outside a container are equal, no further action takes place.

■ COMPRESSIBILITY AND PRESSURE OF GASES

When a gas is compressed to take up a smaller space, the molecules are packed tighter together. Since the molecules in motion (at the same temperature) now strike

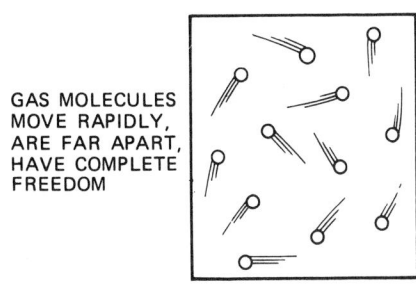
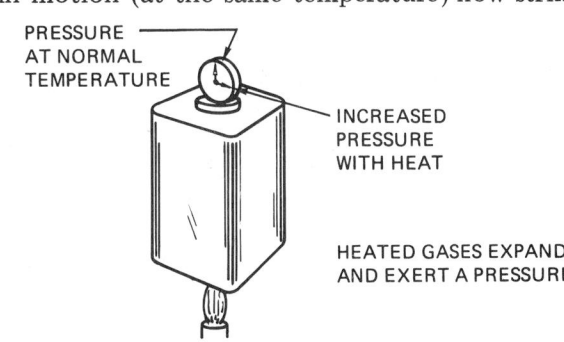

Fig. 5-1

24 Science and Matter — Basic Principles

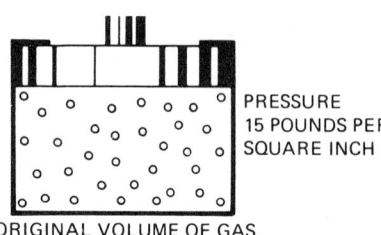 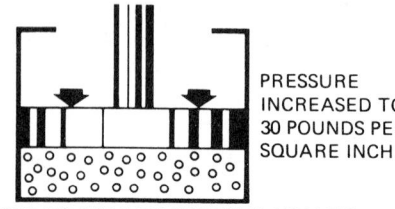

Fig. 5-2

the sides of the container more frequently, the pressure rises. Tests made with gases prove that the pressure of a gas varies inversely with its volume (when the temperature of the gas remains constant). Thus, if a gas is forced (compressed) to occupy a space one-half as large as its original volume and there is no increase in temperature, the pressure is twice the original value, figure 5-2.

Temperature Rise Affects Pressure

The pressure of a confined gas increases as the temperature of the gas rises. This fact is explained by the movement of the gas molecules.

An understanding of pressure is very important. Explosions often result because people overlook the simple facts of the expansion of gases and increased pressures resulting from the application of heat.

Buoyancy of Gases

All gases have weight. Air, which is a mixture of nitrogen, oxygen, and other gases, weighs almost 1 1/4 ounces per cubic foot. The weight of air is taken as a standard against which other gases are compared. It is the weight of air that creates a buoyant force and, at the same time, applies pressure on everything.

Atmospheric Pressure

At sea level there is a pressure of almost fifteen pounds per square inch. This pressure is the weight of air pressing down on objects. The higher an object rises above the earth's surface the less is the pressure on the object. This fact is used by designers, engineers, and others who design guided missiles, rockets, aircraft, and other mechanisms where it is necessary to compensate for changes in pressure. The pressure of the air is called *atmospheric pressure*. This pressure is measured by an instrument known as a *barometer*. The barometer indicates whether the pressure of the atmosphere has increased or decreased from a fixed standard.

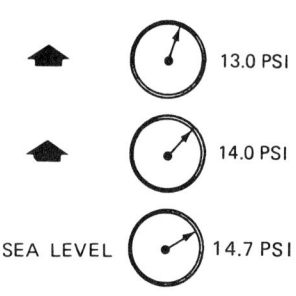

Fig. 5-3 Atmospheric pressure decreases with altitude

■ DENSITY OF GASES

The term *density*, as applied to a gas, means the weight of the gas contained in a unit volume, such as a cubic inch or a cubic foot. The weight of a cubic foot of ammonia gas is 0.045 pound (20.36 grams); for air, the weight is approximately 0.075 pound (34.05 grams); for bottled cooking gas (propane), the weight is 0.117 pound (53.12 grams). Thus, ammonia gas, which is lighter than air, rises because of the buoyant effect of air; propane gas sinks.

■ SPECIFIC WEIGHT OF GASES

Specific weight is a numerical comparison between the weight per volume of a gas and the weight of the same volume of air. A table of *Comparative Densities of Gases Used in Industry* is included in the Appendix. For example, the table shows that the specific weight of ammonia is 0.60, that of air is 1.00, and propane gas is 1.56. Note that the standard specific weight of air is 1.00.

■ INDUSTRIAL APPLICATIONS OF GASES

A knowledge of the specific properties of gases is important to insure that they are handled safely in industry and in other daily uses. Carbon dioxide is often compressed and used in fire extinguishers because of the cooling effect of the gas as the pressure is relieved.

The compressibility of gases is important to welding in that the compressed gases can be transported safely to the job for cutting and fusing operations. Modern food processing includes the packaging of foods in pressure containers. These containers are based on the gas laws and have limitless applications for the packaging of paints, medicines, and cosmetics.

■ CHANGE OF STATE

Changing from a Gas to a Liquid or a Solid State

When the temperature of a gas is decreased to an extremely low temperature, the gas may change to a solid form. Dry ice is compressed carbon dioxide gas. The compressed carbon dioxide is changed to a solid when its temperature is lowered to a point 78 Celsius degrees less than the freezing temperature of water ($0°$ Celsius). Under such conditions, it is possible to change the state of many gases to another form.

Evaporation of Liquids to Form Gases

Many liquids may change their state to become gases. Turpentine, alcohol, gasoline, and paint removers are examples of common liquids that evaporate quickly to form vapors. These liquids are said to be *volatile*. **The gases or vapors of many volatile liquids are explosive, highly flammable, and dangerous when mixed with air. This is one reason why containers of liquids which evaporate into or give off volatile vapors should be sealed and kept away from flame.**

SUMMARY

- Gases expand and become less dense when heated.
- Gas pressure in a closed container increases with an increase of temperature.
- The pressure of a compressed gas increases inversely as its volume is decreased.
- The weight of air creates a buoyant effect.
- The air pressure at sea level is almost 15 pounds per square inch.
- Air pressure decreases as the altitude increases.
- The density of a gas is the weight per volume; the density of air is 0.075 pounds per cubic foot.
- Specific weight is a number which compares the weight per volume of a gas with the weight of the same volume of air. When these quantities are divided (gas by air), the numerical result is the specific weight.
- Volatile liquids evaporate and change into the gaseous state.

ASSIGNMENT UNIT 5 BASIC PROPERTIES OF GASES

■ PRACTICAL PROBLEMS ON PROPERTIES OF GASES

Molecular Motion in Gases

Select the correct word to complete statements 1 to 5.

1. Molecules in gases move freely in (one) (any) direction.
2. Gases (contract) (expand) when heated.
3. The speeds of the molecules in a gas (increase) (decrease) when heated.
4. A (heated) (cooled) gas is lighter than the gas it displaces.
5. A greater pressure is built up in a container as a result of the faster movement of (small) (large) numbers of gas molecules.

Compressibility and Pressure of Gases

For statements 1 to 4, determine which are true (T) and which are false (F).

1. As a gas is compressed, the volume decreases and the pressure increases.
2. The atmospheric pressure at sea level is almost 15 pounds per square inch.
3. Air pressure increases as the distance traveled from the earth's surface increases.
4. A barometer is used to measure temperature.

Density and Specific Weight of Gases

Refer to the table of *Comparative Densities of Gases Used in Industry* in the Appendix.

1. Determine the specific weights of (a) hydrogen, (b) carbon monoxide, (c) oxygen, and (d) butane.

2. Check the table for the density of (a) acetylene, (b) nitrogen, (c) oxygen, and (d) gasoline vapors.

3. Prepare a simple table to show which of the seven different gases in Problems 1 and 2 are (a) less dense than air and (b) more dense than air.

Industrial Applications of Gases

1. List three gases that are lighter than air and give a practical industrial application of each.

2. Name three gases that are heavier than air. Give a common use of each gas.

Change of State

1. Indicate which of the following are volatile liquids.
   a. Motor oil
   b. Ether
   c. Benzine
   d. Carbon tetrachloride
   e. Water
   f. Kerosene
   g. Paints (oil base)
   h. Paints (water base)
   i. Lacquer
   j. Gasoline

2. State two safety precautions that must be observed with volatile fluids.

3. Describe briefly two conditions which produce a change in the state of matter.

Achievement Review of Science and Matter

BASIC PRINCIPLES OF SCIENCE AND MATTER

■ SCIENCE AND THE SCIENTIFIC METHOD

Complete statements 1 to 5 by adding the correct word or phrase.

1. The scientist, inventor, and technologist seek out and apply _____ _____.
2. Scientific knowledge is derived from _____ and _____.
3. _____ is that branch of the physical sciences dealing with the structure and properties of metals.
4. The *scientific method* is a _____ of arriving at fundamental truths.
5. Before a scientific fact is valid it must be _____.
6. Name one new important industrial mechanism or product.
   a. Identify two materials used in its construction.
   b. List at least three different branches of science which contributed to the development or manufacture of one of the materials.

■ THE SCIENCE OF MATTER

Match each term in Column I with the correct description in Column II.

Column I	Column II
1. Porosity	a. The reluctance of a body to change its position of rest, speed, or direction
2. Volume	b. The smallest particle of a substance retaining the original properties
3. Inertia	c. The mass for a unit volume
4. Impenetrability	d. The arrangement or spacing of molecules in a material
	e. The ability of matter to occupy space
	f. The space occupied by a body

For statements 5 to 9, determine which are true (T) and which are false (F).

5. Steel, oil, and air are all in the same state of matter.
6. The arrangement and structure of molecules in a given material determine its physical characteristics.
7. The mass of a body varies while the weight remains constant.
8. The weight of a body is not affected by its density.

9. Density, cohesion, mass, weight, and porosity are some of the properties which are used to identify matter.
10. a. Prepare a table in which hardwood, cork, plastic, and aluminum are listed vertically and the three properties of weight, density, and porosity are listed horizontally.
    b. Rate the properties of each material, using a scale of (1) to indicate the most or heaviest, (2) next, (3) next, and (4) least.

■ BASIC PROPERTIES OF SOLIDS

Select the correct word or words to complete statements 1 to 5.

1. When one material cuts another, the material doing the cutting is the (harder) (softer) of the two.
2. (Conductivity) (Elasticity) (Fusibility) is the ability of matter to change state and be welded together.
3. A (ductile) (hardened) (tough) metal may be drawn easily into a different shape.
4. A (hardened) (malleable) metal part has the greater ability to resist scratching.
5. (Toughness) (Elasticity) is the tendency of a material to return to its original shape when a force is removed.

Identify which of the properties (a to h) is the main property of each material (6 to 15).

6. Industrial diamond
7. Welding rod
8. Spring steel
9. Glass (ordinary)
10. Rubber
11. Abrasive grains
12. Aluminum wire
13. Asbestos
14. Silver
15. Wood (oak)

a. Hard
b. Tough
c. Malleable
d. Elastic
e. Ductile
f. Machinable
g. Fusible
h. Heat-resistant

■ BASIC PROPERTIES OF LIQUIDS

Indicate the term which each statement (1 to 6) describes.

1. _____ is that property of liquids to wet the surface of a solid.
2. _____ refers to the ability of the molecules of a liquid to resist change.
3. _____ is the ability of a liquid to support an object by exerting a force.
4. _____ is the tendency of the surface of a liquid to become as small and as tough as possible.
5. _____ depends on cohesion, adhesion, and surface tension.

30   Science and Matter — Basic Principles

6. _____ is the ratio of a given volume of a material to an equal volume of water.
7. Name three common metals that are heavier than water.
8. Identify three common materials that float in oil.

■ BASIC PROPERTIES OF GASES

Complete statements 1 to 6 by adding the correct word or phrase.

1. When heated, gases _____ and become _____.
2. The gas pressure in a closed container _____ as the temperature increases.
3. The ratio of the weight per volume of gas to air is the _____ of the gas.
4. The _____ of a gas is the weight per volume.
5. A _____ liquid is one which evaporates and changes to the gaseous state.
6. _____ refers to the pressure of the air pushing down on an object.
7. a. List two industrial gases that are lighter than air.
   b. Give a practical application of each one.
8. a. Name two common gases that are lighter than air.
   b. Give two examples of where each one is used.

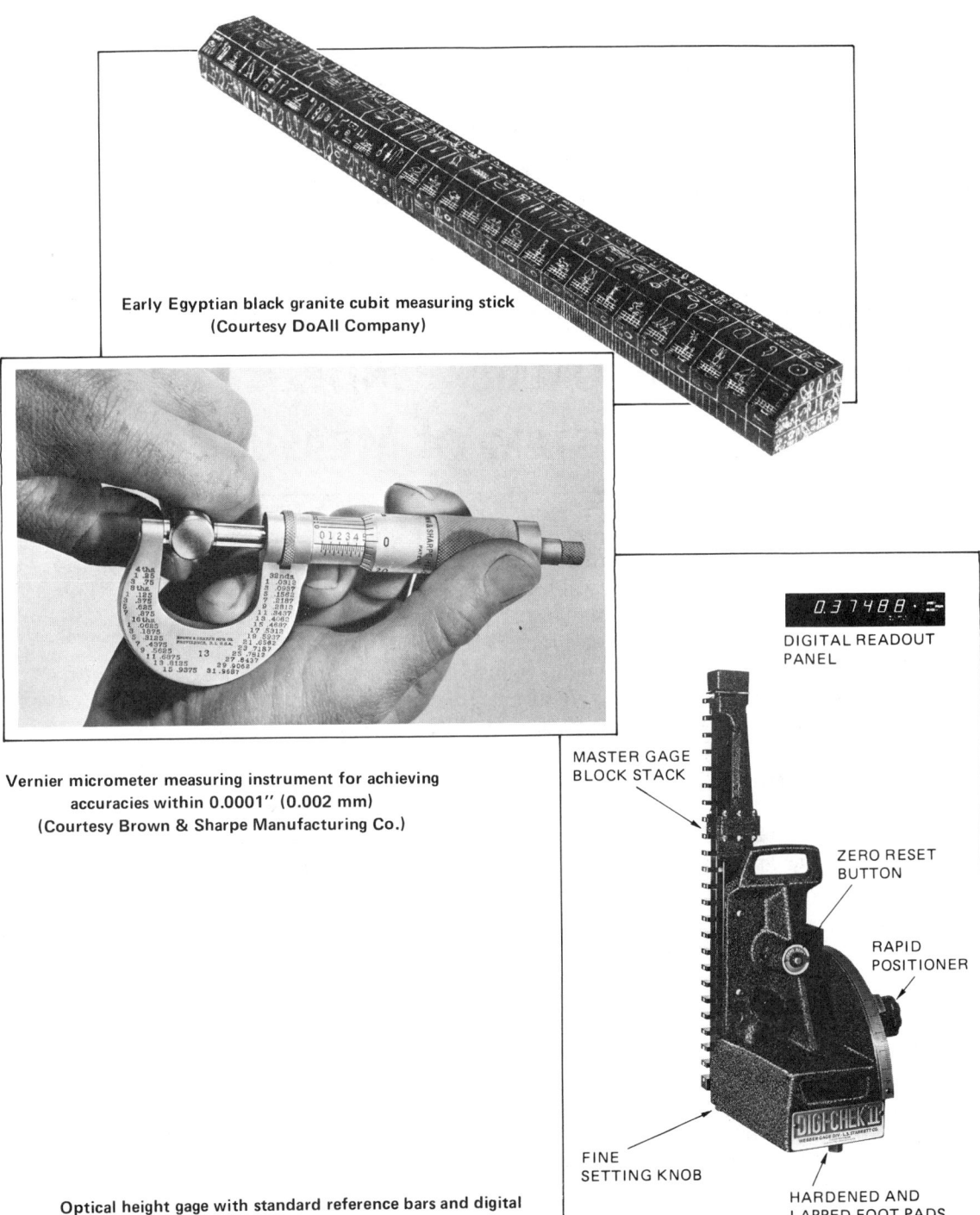

Early Egyptian black granite cubit measuring stick
(Courtesy DoAll Company)

Vernier micrometer measuring instrument for achieving
accuracies within 0.0001" (0.002 mm)
(Courtesy Brown & Sharpe Manufacturing Co.)

Optical height gage with standard reference bars and digital
readout for achieving accuracies within 0.0001" (0.002 mm)
(Courtesy Webber Gage Division, L. S. Starrett Co.)

Early and evolving examples of
linear measurement instruments

Section 2
Systems of Measurement in Science

Unit 6 Customary United States and British System of Measurement

Scientific progress depends upon the degree to which extremely fine particles of matter can be identified and measured. When a measurement is taken directly with the use of tools, instruments, or other calibrated measuring devices, it is known as *direct measurement*. When the measurement must be determined using a formula (computation), the measurement is called an *indirect* or *computed measurement*.

The most important measurements considered in this section deal with length, mass, and time. Measurements related to heat, light, and electrical energy are treated in later sections. These measurements are applied as each new scientific principle is studied.

Measurements are used internationally as a technical language in which precise relationships are stated for physical quantities. There are two parts to every measurement: (1) quantity and (2) a designation of the nature of the quantity. The magnitude of a physical quantity is specified by a numerical value and a unit.

■ STANDARDS AS CONSTANT MEASUREMENTS

The need for a standard like the Customary inch or metric meter units of measurement) to be absolutely constant in magnitude has caused scientists, engineers, and other measurement specialists to fix each value to an invariant quantity in nature. An *invariant quantity in nature* is one that is regarded as a constant (without variation) under all circumstances. In other words, it is unchanging with time.

Until recently, the standard measurement of length was based on a specific platinum-iridium bar kept in Sevres, France. Today the meter is specified as 1 650 763.73 wavelengths of one of the spectral lines of a krypton isotope (^{86}Kr). Established in this manner, a precise meter measurement can be reproduced in any laboratory in the world. Having the standard meter (m) measurement, the Customary inch (in) unit may be precisely established as $\frac{1}{39.37}$ of a meter (m), or 0.0254 m. The standard meter in the SI Metric system is 39.37 inches.

■ MEASUREMENT OF LENGTH

The measurement of length is often called *linear measure* or the measure of straight line distances between two points, lines, or surfaces. These measurements may be expressed in either the metric system or the British system. Of the two systems, the British system of linear measure is more widely used in the United States. While the foot is the standard unit of length in the British system, the inch is the smallest unit of measure in the U.S. Customary measurement system.

The tools most commonly used in the laboratory for measuring length are the yardstick (meter stick in the metric system), steel rule, measuring tape, micrometer, vernier caliper, vernier micrometer, and gage blocks. The measurements taken with these tools vary in accuracy from about 1/32 inch with the yardstick to 0.0001 inch with the vernier micrometer.

Yardsticks, Tapes, and Rules

The inch, which is the smallest unit of linear measure in the British system, is subdivided into smaller fractional parts. The fractional divisions of an inch which are most commonly used on yardsticks and steel tapes represent halves, quarters, eighths, sixteenths, and thirty-seconds of an inch.

The steel rule has a greater degree of accuracy than either the yardstick or the steel tape. There are two standard methods of graduating steel rules: (1) into fractional divisions, and (2) into decimal divisions, figure 6-1.

The fractional divisions represent halves, quarters, eighths, sixteenths, thirty-seconds, and sixty-fourths of an inch. When smaller units of measure are needed, steel rules graduated in halves, tenths, fiftieths and hundredths of an inch are used. The decimal rule is convenient to use for accurate measurements because the possibility of error in changing from common fractions to decimals is reduced.

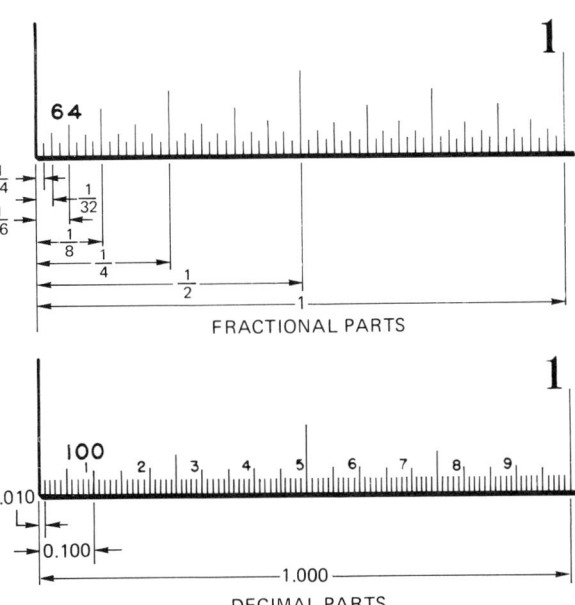

Fig. 6-1 Enlarged views of subdivisions of an inch

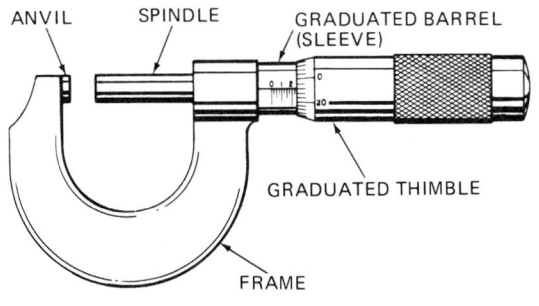

Fig. 6-2 The micrometer

■ PRECISION MEASUREMENTS

The Micrometer (0.001 inch). Laboratory measurements of length are often

34 Systems of Measurement in Science

required to an accuracy of one ten-thousandth part of an inch or more. For most practical purposes, an instrument called a *micrometer* is used, figure 6-2.

Direct readings on the standard micrometer are obtained by turning a graduated thimble which moves a spindle toward or away from a fixed anvil. The spindle is carefully advanced to the workpiece until the space between the spindle and the work is just large enough to permit the micrometer to be moved over the work. The measurement is read directly from the graduations on the barrel and the thimble. The visible vertical sleeve graduation (in multiples of 0.025" increments) and the 0.001" increment value of the thimble graduation at the index line are added.

■ MEASUREMENT OF AREA AND VOLUME

The measurement of the space occupied by a body having the three linear dimensions of length, height, and depth is expressed as cubic or volume measure. When the three dimensions can be taken directly from a solid shape such as a cube, a rectangular solid, or a cylinder, the volume can be computed. The volume of irregularly shaped objects can be determined by the displacement method which will be described later.

Concept of Square Measure

Common activities of daily living and work often require the determination of the space occupied by a flat surface. The space occupied by a rectangular plot of land is the product of its length and width. Mathematically, this computed quantity is called the area of the object and is expressed in square units of the same kind as the linear units.

One square inch is the area occupied by a square object which is one linear inch long and one linear inch high, figure 6-3. One square foot is the area occupied by a square object which is 12 linear inches long and 12 linear inches high.

One square yard is the area of a square object which is 36 inches long and 36 inches high. The area of such a square is also 1296 square inches or 9 square feet, figure 6-4. The units in which the area is expressed depend on whether the dimensions of the length and height are given in inches, feet, or yards. Figures 6-4 and 6-5 compare the square inch, square foot, and square yard.

Area measure is used to compute the size of square, rectangular, round, and irregularly shaped surfaces. A solid is formed when a third dimension of depth is added to these surfaces.

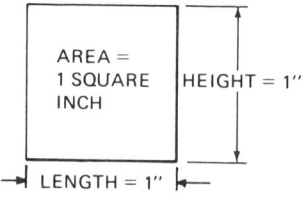

Fig. 6-3

Table of Surface Measure	
Unit of Surface Measure =	1 sq in
144 sq in =	1 sq ft
9 sq ft =	1 sq yd

Fig. 6-4

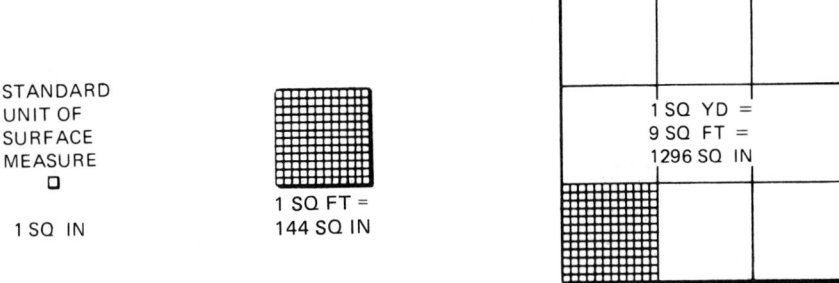

Fig. 6-5 Comparison of units of area measure

Concept of Volume Measure

Volume measure refers to the measurement of the space occupied by a body having three linear dimensions of length, height, and depth. The principles of volume measure are applied in this unit to common shapes such as the cube, the rectangular solid, the cylinder, and various combinations of these shapes, and to objects that are irregularly shaped. For these objects, the volume may be determined more accurately by methods other than computation.

Cubes

The volume of a solid is the product of three linear dimensions. Each dimension must be in the same unit before they can be multiplied. For example, if the length, height, and depth of a cube are given in feet, then the volume will be in cubic feet.

Figure 6-6 compares the volume of cubes one inch, one foot, and one yard on a side respectively. Note that one cubic foot = 1728 cubic inches and one cubic yard =

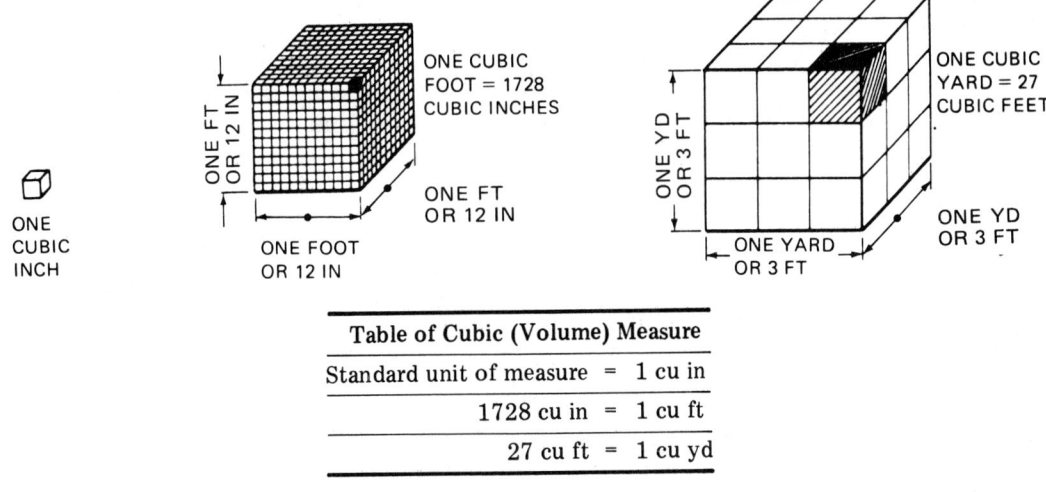

Table of Cubic (Volume) Measure	
Standard unit of measure =	1 cu in
1728 cu in =	1 cu ft
27 cu ft =	1 cu yd

Fig. 6-6 Comparison of units of volume measure

27 cubic feet. The measurement of the volume of a cube is equal to its length × depth × height.

Rectangular Solids

When the surface of a flat rectangle is extended in a third direction, a rectangular solid is formed, figure 6-7. The space occupied by this solid is measured in terms of the number of cubic units which it contains. The volume measure of a rectangular solid is equal to the product of the length, depth and height.

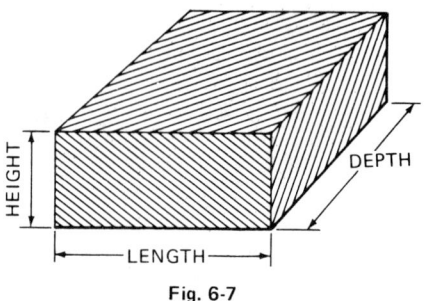

Fig. 6-7

Cylindrical Solids

The volume of a cylinder is a measure of the cubic units which it contains. The volume is equal to the area of the circular base multiplied by the length or height of the cylinder, figure 6-8.

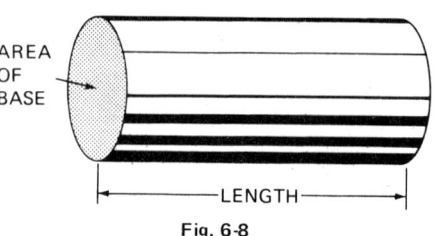

Fig. 6-8

Displacement Method of Measuring Volume

There are two types of irregularly shaped objects. The first type consists of a combination of regular solid shapes. For this object, the volume can be determined easily by computation. The second type of object may be so irregular in shape (a lump of coal, for example) that its volume must be determined by a simpler and more accurate method than computation.

The *displacement method* is based on the principle of the impenetrability of matter. This principle states that two bodies cannot occupy the same space at the same time. As a result, when a solid is immersed in a liquid, the volume of the liquid is noted before and after the body is placed into it. In this manner, the volume of the body or solid can be determined quickly.

In figure 6-9, the volume of water in the container is 30 cubic inches. An irregularly shaped metal part is placed in the container. The new volume of the water and the object immersed in it is equal to 50 cubic inches. The volume of the metal part can be determined by subtracting the original volume of 30 cubic inches from the new volume of 50 cubic inches. Thus, the volume of the metal part is 20 cubic inches.

An object may be bulky and may not fit conveniently into a graduated vessel. To measure the volume of this object, a tank fitted with an overflow spout is filled with water to the spout level. As the object is lowered into the tank, the overflow of water is collected in a separate container and measured. The volume of the displaced water is equal to the volume of the object.

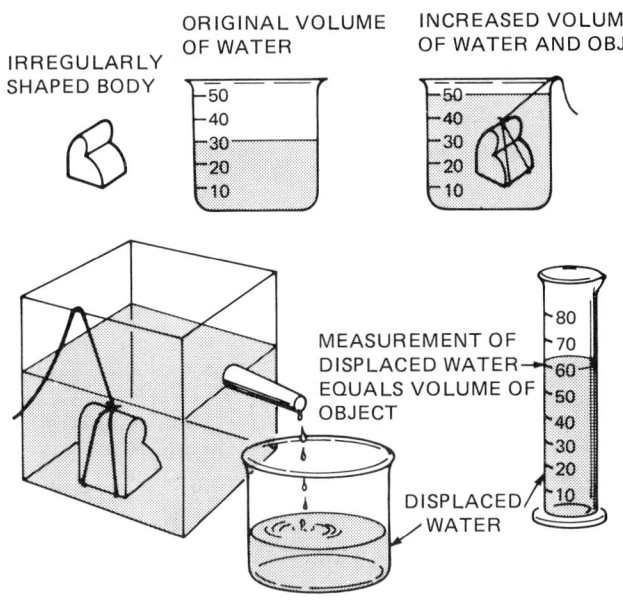

Table of Liquid Measure	
2 cups	= 1 pint (pt)
2 pints (pt)	= 1 quart (qt)
4 quarts (qt)	= 1 gallon (gal)
	= 231 cu in
31 1/2 gallons	= 1 standard barrel (bbl)

Fig. 6-10

Fig. 6-9

Changing Units of Volume Measure to Liquid Measure

The volume of a solid usually is given in terms of cubic units of measure. A cube one inch on a side has a volume of one cubic inch. Liquids, on the other hand, are measured in the English system by liquid measure. The standard units in this system are the cup, pint, quart, gallon, and barrel, figure 6-10.

By law, the United States gallon contains 231 cubic inches of liquid. This relationship makes it possible to change a unit from cubic measure (inch) to volume measure (gallon) by dividing by 231, or from volume measure to cubic measure by multiplying by 231.

■ MEASUREMENT OF WEIGHT AND TIME

Weight of a Body

The basic units of weight in the English system of measurement are the ounce, pound, hundredweight, and ton. Small objects and mechanisms may be weighed directly with some form of spring balance. The weight of simple shapes can be computed by multiplying the volume of the shape by the density of the material.

The weight of a body may vary when it is moved from place to place because of variations in its distance from the earth's center. However, for all practical purposes, a spring balance is accurate enough for the measurement of weight.

38 Systems of Measurement in Science

Measurement of Time

The units of time measurement are standardized around the world. The standard day is the interval during which the earth rotates once around its axis with respect to the sun. The day is divided into 24 parts called hours. The hour is divided into 60 parts (minutes) and each minute is divided into 60 parts (seconds).

The second in this system of standard time units is 1/86 400th of the mean solar day. The solar day represents the average time required for a single rotation of the earth around its axis in relation to the sun.

SUMMARY

- The measurement of any quantity requires a comparison with a standard quantity (unit) of the same kind.
- Common measuring tools such as rules and tapes are used to take measurements of length to an accuracy of 1/100 inch. Precision measuring instruments yield linear measurements with accuracies ranging from one-thousandth to two-millionths of an inch.
- The displacement method of measuring volume is widely used for irregularly shaped solids. The volume of liquid displaced when an object is immersed is equal to the volume or mass of the object.
- Tables of comparison among metric, English, and SI units of measure furnish values which may be used to convert from one system to another.
- The volume of a body may be computed in both the metric system and the English system. The volume expressed in linear measure can be converted to a unit of liquid measure.
- Conversion is the process of determining a given quantity of a certain unit in terms of another unit of the same kind.

ASSIGNMENT UNIT 6 CUSTOMARY UNITED STATES AND BRITISH SYSTEM OF MEASUREMENT

■ PRACTICAL PROBLEMS WITH CUSTOMARY UNITED STATES AND BRITISH UNITS OF MEASUREMENT

Nonprecision Linear Measurements with Yardstick, Tape, and Rule

1. Determine the reading of each measurement indicated on the yardstick in figure 6-11.

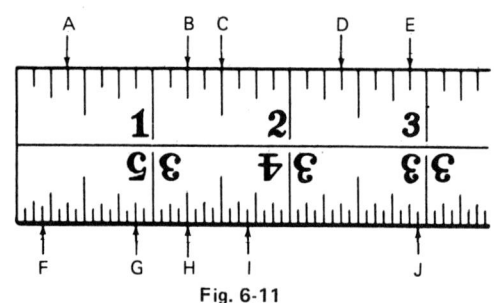

Fig. 6-11

2. Determine the reading of each measurement indicated on the steel rules in figure 6-12. The rule on the left is graduated in decimal parts; the rule on the right is graduated in fractional parts of an inch.

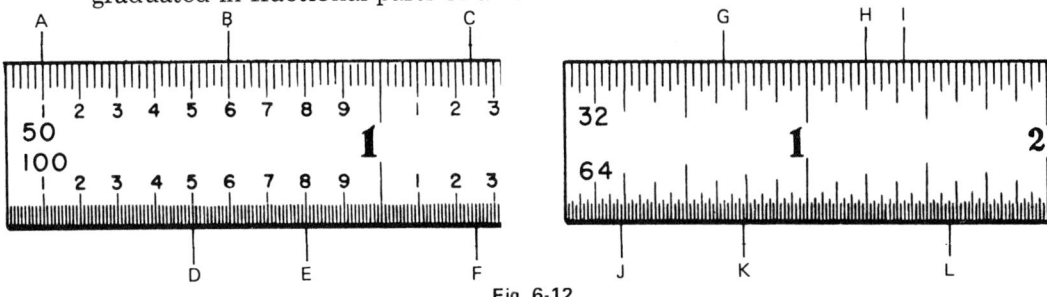

Fig. 6-12

Precision Linear Measurements with the Standard Micrometer

1. Determine the linear dimensions indicated at A, B, C, D, and E on the standard micrometer in figure 6-13.

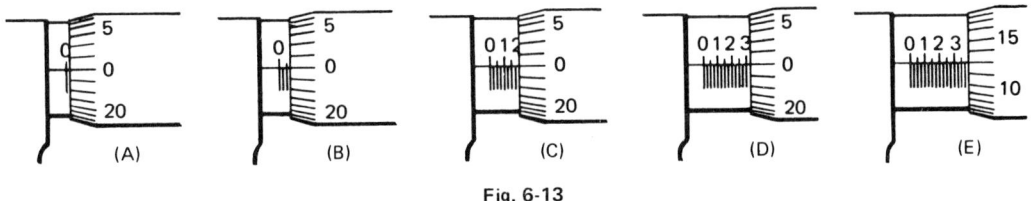

Fig. 6-13

Volumes of Solids (British System)

1. Determine the volumes of cubes A, B, and C and the rectangular solids D, E, and F in figure 6-14.

	A	B	C	D	E	F
Length	10″	2 1/2″	3 1/2″	5″	2 ft	1 yd
Depth	10″	2 1/2″	3 1/2″	2″	1 ft	1 yd 1 ft
Height	10″	2 1/2″	3 1/2″	4″	8 in	1 yd 8 in

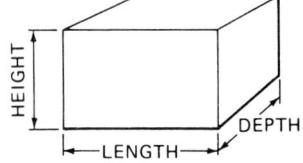

Fig. 6-14

40 Systems of Measurement in Science

2. In figure 6-15, find the volumes of the cylinders A, B, and C and the cored cylinders D and E, correct to the two decimal places.

	A	B	C
Diameter	2"	6"	
Radius			12.2"
Length	4"	4 1/2"	3.4"

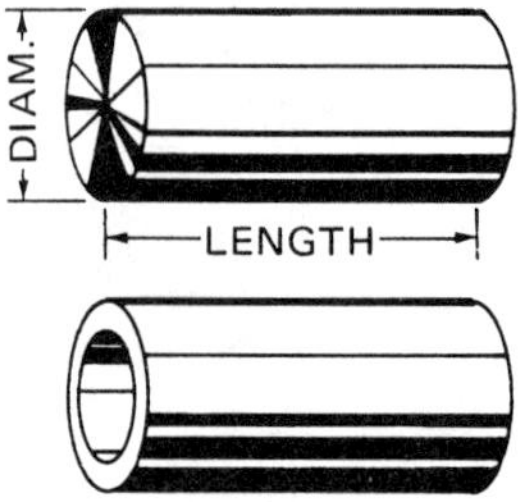

	D	E
Outside Diameter	2"	5"
Inside Diameter	1"	3"
Length	10"	1'-4"

Fig. 6-15

Changing Units of Volume Measure to Liquid Measure (British System)

1. Change the quantities given in the table from one unit of liquid measure to the one stated in each case.

A	12 qt to pints		E	2 bbl to gallons
B	3 gal to quarts		F	3 1/2 bbl to quarts
C	2 gal to pints		G	3 gal 5 qt to quarts
D	8 pt to gills		H	6 3/4 pt to pints and gills

Unit 7 Metrication: SI Metric Units of Measure

There is a slow but persistent effort toward the adoption of the International System of Units (SI) by the United States and other nations which use measurement standards based on the British gravitational system and other metric systems. The SI evolved from a consortium of over fifty industrialized nations in the International Standards Organization (ISO). The term *metrication* is used during these changing times in connection with any program or process of conversion to the International System of Units (SI).

■ ORGANIZATION AND FUNCTIONS OF THE GOVERNMENT AND THE PRIVATE SECTOR IN METRICATION

The Metric Conversion Act

In December of 1975, the President of the United States signed the Metric Conversion Act of 1975. This event gives official federal sanction to a movement which has been gaining in acceptance for several years. The Metric Conversion Act establishes a United States Metric Board which is charged with the tasks of "...devising and carrying out a broad program of planning, coordination, and public education, consistent with other national policy and interests, with the aim of implementing the policy set forth in the Act."

The United States Metric Board works closely with responsible organizations which are involved in implementing metrication in the United States.

The American National Standards Institute (ANSI)

The American National Standards Institute (ANSI) represents the United States at the International Standards Organization. In this role, the ANSI is the national coordinating agency that is involved in the compromises which must be made if acceptable international standards of weights and measures are to be developed. The ANSI is a voluntary national organization which serves all interests and segments of society. Its functions are advisory and it operates under the principles of *nonlegislative consensus*. The ANSI provides the nucleus for planning and coordinating activities for organizations and industries converting to SI units of measurement.

■ BASE TEN SCIENTIFIC NOTATION SYSTEM

The International System of Units (SI) uses a *base ten scientific notation system* which simplifies the designation of quantities and mathematical processes and reduces errors in calculations. Part A of table 7-1 shows standard prefixes and symbols and the equivalent value of each prefix as a power of ten. The values for the positive powers of

Prefix	Symbol	Value as Power of Ten	Multiplication Factor
deka	da	10	10
hecto	h	10^2	100
kilo	k	10^3	1 000
mega	M	10^6	1 000 000
giga	G	10^9	1 000 000 000
tera	T	10^{12}	1 000 000 000 000

(A)

Prefix	Symbol	Value as Power of Ten	Multiplication Factor
deci	d	10^{-1}	0.1
centi	c	10^{-2}	0.01
milli	m	10^{-3}	0.001
micro	μ	10^{-6}	0.000 001
nano	n	10^{-9}	0.000 000 001
pico	p	10^{-12}	0.000 000 000 001
femto	f	10^{-15}	0.000 000 000 000 001
atto	a	10^{-18}	0.000 000 000 000 000 001

(B)

Table 7-1 SI unit prefixes, symbols, and power of ten multiple and submultiple values

ten are expressed by an exponent. For instance 10^2 indicates a quantitative value of 10 · 10 or 100; 10^3 = 10 · 10 · 10 or 1000.

Submultiples of the base units are illustrated in Part B of table 7-1. Each prefix is identified by a different symbol. The value of each prefix is shown as a negative power of ten. A negative exponent indicates that the value is less than (1). If there is no negative sign (−) in front of the exponent, then the exponent is positive.

Performing Mathematical Processes in the Base Ten System

There are two basic steps in solving multiplication and division problems using values in whole numbers or decimals in the base ten system. First, the whole numbers or decimals are multiplied. Secondly, the exponents are added or subtracted depending on the (+) or (−) value of the exponent. The value of the answer is expressed as a positive or negative quantity.

Example 1. Multiply 4.4 to the 10^{-7} power by 3.1 to the 10^{-9} power.
- Multiply the decimal values (4.4) · (3.1) = 13.64
- Add the exponents $(10^{-7}) + (10^{-9}) = 10^{-16}$
  Answer: **13.64 (10^{-16})**

All scientific problems deal with quantities and units of measurement. Thus, each solution must include an identification of the quantity and unit of measurement.

Example 2. Multiply 4.4 to the 10^{-7} power by 3.1 to the 10^9 power.
- Multiply the decimal values $(4.4) \cdot (3.1) = 13.64$
- Add the exponents $(10^{-7}) + (10^9) = 10^2$

Answer: **13.64 (10^2) or 1364**

For scientific problems involving the division of whole numbers or decimal values which are expressed in powers of ten, the division of the values is completed first. Then the sign of the exponent in the denominator is changed and added to the exponent in the numerator.

Example 3. Divide 4.4 (10^3) by 2.2 (10^{-6})
- Divide 4.4 by 2.2 = 2.0
- Change the sign of the exponent in the denominator (10^{-6}) to (10^6)
- Add the values of the exponents $(10^3) + (10^6) = 10^9$

Answer: **2 · (10^9)**

This is a simplified method of writing the value 2 000 000 000.

Example 4. Divide $4.4 \cdot 10^3$ by $2.2 \cdot 10^4$
- Divide 4.4 by 2.2 = 2.0
- Change the sign of the exponent (10^4) to (10^{-4}) and add it to the numerator (10^3)

$(10^{-4}) + (10^3) = 10^{-1}$

Answer: **2.0 · 10^{-1} = 0.2**

■ ADVANTAGES OF SI METRICS

SI is recognized in all languages as the abbreviation of the International System of Units. The full title and abbreviation were accepted in 1960 by 36 countries, including the United States. Some of the more important advantages of SI are as follows:

- SI provides only one base unit for each physical quantity: meter (length), kilogram (mass), second (time), kelvin (temperature), ampere (electrical current), candela (luminous intensity), and mole (substance of a system). Additional units are derived from these base units. The derived units are defined by simple equations.
- SI provides a unique and well-defined set of symbols and abbreviations, each of which relates to a specific phenomenon or condition.
- SI retains the decimal relationship between multiples (+) and submultiples (−) of the base unit for each physical quantity. Prefixes are used to simplify long multiple and submultiple values.

44 Systems of Measurement in Science

- SI is a *coherent system*. It has seven base units with established names, symbols, and precise definitions. Coherence also means that a product or quotient of any two quantities is a unit of the resulting quantity. For example, a unit area is the product of a unit of length and another unit of length. In a coherent system where the meter is the unit of length, the square meter is the unit of area.

- The base units of SI are accurately defined in terms of physical measurements. With the exception of the kilogram mass that is preserved at the International Bureau of Weights and Measures, the measurement for each base unit can be produced in laboratories.

- Other units that are not a part of SI, but are associated with it, are related to the base units of the system by powers of ten.

■ SEVEN BASE UNITS OF SI METRICS

The term *SI metrics* is used in this text as a modification of the accepted SI abbreviation of the International System of Units. Although there is variation among the nations on the spelling of scientific terms, the spelling of the SI terms in this and subsequent units follows the common usage in the United States.

There are seven *base units*, two *supplementary units*, and other additional units that are *derived* from the base units in SI metrics.

Unit of Length

The unit of length in SI metrics is the meter (m). A *meter* is defined as a length that is equal to the distance travelled by light in a vacuum during 1/299,792,458 of a

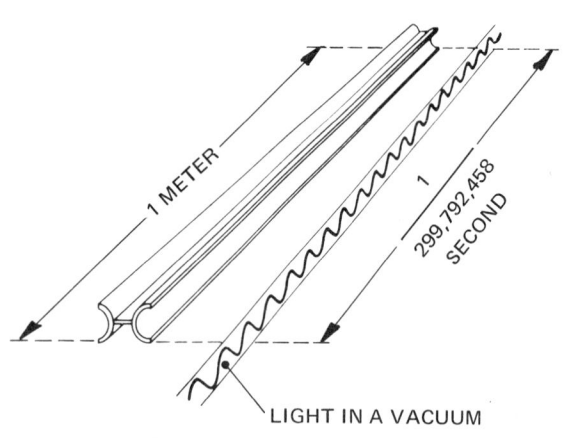

Fig. 7-1

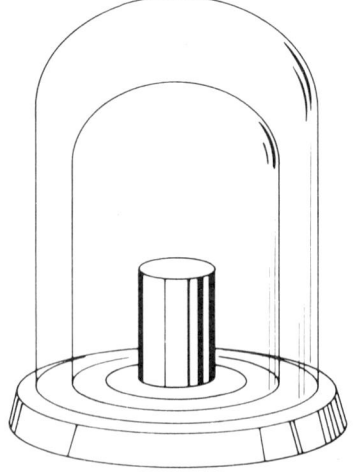

Fig. 7-2 U.S. kilogram mass prototype #20

second, figure 7-1. A definition of this type allows laboratory technologists and scientists anywhere in the world to duplicate the meter with extreme accuracy. Linear measurements are generally given in SI metrics in kilometers, meters, centimeters, and millimeters.

Unit of Mass

The unit of mass is the kilogram (kg), figure 7-2. The *kilogram* is equal to 2.2 pounds. Thus, the mass of a body weighing 220 pounds is 100 kilograms. Another body weighing 200 pounds has a mass of 90.909 kg. The gram is also widely used as an SI metric unit of mass. One *gram* (g) represents the mass of one cubic centimeter (1 cm^3) of water at 4°C.

Unit of Time

The duration, or time interval of one second, is still 1/86 400 of the mean solar day.

However, scientists today need time units divided as finely as billionths of a second. This need for smaller and smaller units of time measurement led to a new standard for the second based on the periodic time intervals stimulated in a beam of cesium (^{133}Cs) atoms, figure 7-3. The *cesium-clock system* defines the standard second as the time of 9 192 631 770 vibrations in a beam of ^{133}Cs atoms. An important property of this natural time unit is the fact that it is reproducible anywhere in the world.

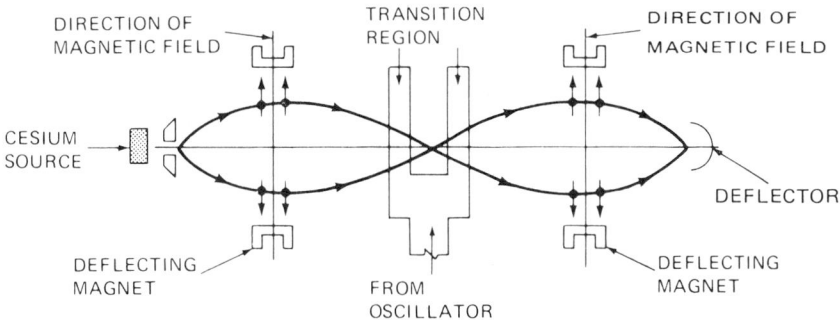

Fig. 7-3

Unit of Electrical Current

The SI unit of electrical current is the ampere (A). The *ampere* is defined as "... that constant which, if maintained in two straight parallel conductors of indefinite length (of negligible cross section) and placed one meter apart in a vacuum, will produce a force between these conductors equal to 2×10^{-7} newton/meter of length," figure 7-4.

An ampere is the rate of flow of one coulomb of electrical charge passing a given point in one second.

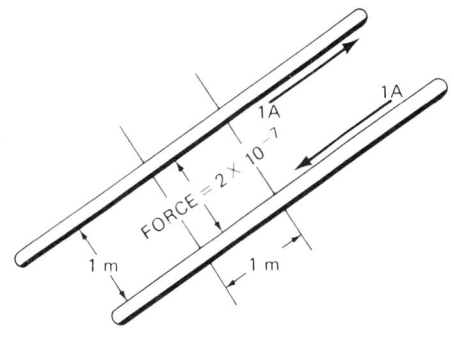

Fig. 7-4

Unit of Temperature

The unit of thermodynamic temperature in SI is the kelvin (K). The *kelvin* is equal to 1/273.16th of the thermodynamic temperature of the triple point of water. On the Kelvin scale, water boils at 373.15 K degrees and freezes at 273.15 K degrees.

A more practical and commonly used unit of temperature measurement is the *Celsius degree* (C°). Water freezes at 0° and boils at 100° on the Celsius scale (formerly called the centigrade scale). To convert from a °C to a kelvin degree, add 273.15 to the Celsius measurement: K = °C + 273.15.

Unit of Luminous Intensity

The unit of luminous intensity is the candela (cd). The *candela* represents the luminous intensity (in the perpendicular direction) of a surface 1/600 000th of a square meter of a blackbody at the temperature of freezing platinum under a pressure of 101 325 newtons per square meter, figure 7-5.

Unit of Substance of a System

The seventh base unit in SI metrics is the mole (mol). The *mole* represents the amount of substance of a system which contains as many elementary entities as there are atoms in 0.012 kilogram of carbon-12. Elementary entities such as atoms, molecules, ions, electrons, and other particles must be specified when the mole is used.

■ SUPPLEMENTARY UNITS OF ANGULAR MEASURE

SI metrics has two supplementary units which deal with two types of angles: plane and solid. The *plane angle* or *radian* (rad), figure 7-6, is the unit of measure of an angle which has its vertex at the center of a circle and is subtended by an arc that is equal in length to the radius.

1 rad = 180°/π = 57.259 78°

The *solid angle* or *steradian* (sr), figure 7-7, has its vertex at the center of a

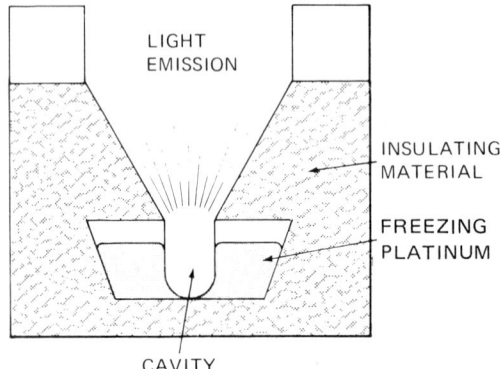

Fig. 7-5

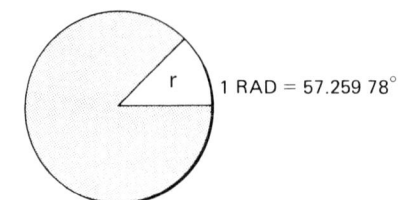

Fig. 7-6 Plane angle

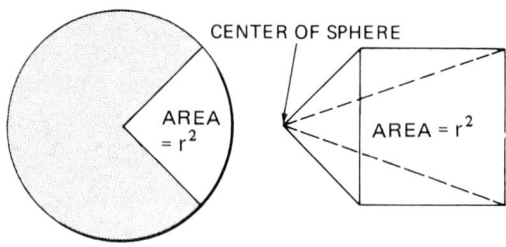

Fig. 7-7 Solid angle

sphere and encloses an area of the spherical surface. This area is equal to that of a square whose sides are equal in length to the radius of the sphere.

■ DERIVED UNITS OF MEASURE

Derived Area and Volume Units

The following units of area are derived from the meter as a base unit.

- The *square meter* (m^2)
- The *hectare* (h) which is equal to 10 000 m^2
- The *cubic meter* (m^3) which is used in volume and capacity measurements
- The *liter* (L) which equals 1000 cm^3. The liter is an acceptable unit of volume measure in SI because it can be defined in terms of an SI base unit.

Units Derived from the Unit of Mass

Units of force, work (energy), power, and pressure are all derived from the base unit of mass.

Force. *Force* is measured in newtons (N). A *newton* is equal to (mass) · (acceleration) or kilogram · meter per second squared (N = kg · m/s^2).

> A force of one newton applied to a one-kilogram mass produces an acceleration of one meter per second squared.

Work. *Work* is measured in joules (J). A *joule*

$$= \text{force} \cdot \text{distance}$$
$$= \text{newtons} \cdot \text{meters}.$$
$$J = (N) \cdot (m)$$
$$= N \cdot m$$

One joule of work is produced when one newton is applied through a distance of one meter.

Power. *Power* is measured in watts (W). A *watt* is equal to one joule per second or newtons (N) · meters per second (m/s). W = J/s = N · m/s.

One watt of power represents one joule of work completed in one second.

Pressure. *Pressure* is measured in pascals (Pa). A *pascal* is equal to one newton per meter squared (Pa = N/m^2).

> One pascal is equal to the force of one newton applied to an area of one square meter.

Units Derived from the Base Unit of Time (Second)

The hertz (Hz), velocity (v), and acceleration (a) are three units that are derived from the SI base unit of time, the second.

Frequency. The *hertz* (Hz) measures frequency or the number of cycles per second.

Velocity. *Velocity* (v) refers to distance divided by time or meters per second.

Acceleration. *Acceleration* (a) is the derived unit which expresses the rate of change in velocity. Acceleration is measured in meters per second per second or meters per second squared (m/s^2).

Units Derived from the Base Electrical Unit (Ampere)

	Measurement	Derived Unit of Measurement	Formula
Base Electrical Unit: Ampere (A)	Electrical potential	volt (V)	V = W/A
	Electrical resistance	ohm (Ω)	Ω = V/A
	Electrical capacitance	farad (F)	F = A · (s/V)
	Quantity of electricity	coulomb (C)	C = A · s
	Electrical inductance	henry (H)	H = Wb/A
	Magnetic flux	weber (Wb)	Wb = V · s

Table 7-2

Units Derived from the Base Unit of Luminous Intensity

	Derived Units	Characteristics	Formula
Base Unit of Luminous Intensity: candela (cd)	Illumination, lux (lx)	Luminous intensity given by a luminous flux of one lumen per square meter.	lx = lm/m^2
	Luminous flux, lumen (lm)	Luminous flux emitted in a solid angle of one steradian by a point source with a uniform intensity of one.	lm = cd · sr
NOTE: 1 watt of electrical power = 17 lumens			

Table 7-3

SI Metric Base Units, Supplementary Units, and Derived Units

Quantity	Base Units	Symbol
length (l)	meter	m
mass (m)	kilogram	kg
time (t)	second	s
electric current (I)	ampere	A
thermodynamic temperature (T)	kelvin	K
luminous intensity (I)	candela	cd
substance	mole	mol
Quantity	Supplementary Units	Symbol
plane angle	radian	rad
solid angle	steradian	sr

Table 7-4 Common SI metric units and symbols

Quantity	Derived Units (Selected)	Formula
acceleration (a)	meter per second squared	m/s^2
area (A)	square meter	m^2
density (D)	kilogram per cubic meter	kg/m^3
electric capacitance (C)	farad (F)	$A \cdot s/V$
electric charge (Q) (quantity of electricity)	coulomb (C)	$A \cdot s$
Energy (E)	joule (J)	$N \cdot m$
force (F)	newton (N)	$kg \cdot m/s^2$
frequency (f)	hertz (Hz)	(cycle)/s
magnetic flux density (B)	tesla (T)	Wb/m^2
magnetomotive force	ampere (A)	
power (P)	watt (W)	J/s
pressure (p)	newton per square meter	N/m^2
thermal conductivity	watt per meter · kelvin	$W/m \cdot K$
velocity (v)	meter per second	m/s
work (W)	joule (J)	$N \cdot m$

Table 7-5 Common SI metric units, symbols, and formulas: Derived Units

GUIDELINES FOR WRITING AND USING SI METRIC UNITS

Basic rules govern the method of expressing values and units in the SI Metric system. The following rules promote uniformity in writing and accuracy in interpretation of the units.

- Base units, supplementary units, derived units, and combinations of these units, with multiple and submultiple quantities, are used.
- Insignificant digits and decimals are eliminated.
- Powers of ten are used for computation.
- Prefixes express the order of magnitude of a value. For example, 56.5 km is a precise definition of a specific measurement.
- Prefixes indicating values of 1000 are preferred in the SI Metric system. For example: milli (10^{-3}), kilo (10^3), and mega (10^6).
- Symbols are capitalized in the SI Metric system when they are derived from a proper name: Henry (H), Siemens (S), and Testa (T).
- The numerical prefixes of tera (T), giga (G), and mega (M) are also capitalized.
- The SI Metric system symbols are written in singular form when abbreviated. For instance, ten meters is written as 10 m; 16.2 kilograms as 16.2 kg.
- Number values are written in groups of three digits without commas. For example, 2755932.6 Hz is written 2 755 932.6 Hz. Similarly, the value of 8.37250 W is written 8.372 50 W. However, the space is not recommended for four-digit numbers, unless such numbers are grouped in a column with numbers of five digits or more.
- A zero is used before a decimal quantity: 0.2 kg.
- A heavy dot (·) is used in mathematical equations to indicate multiplication. A slash line (/) indicates division.
- Values may be rounded off to a stated degree of accuracy after all calculations are complete. The process of rounding off may be started before this stage if the conversion factors contain more digits in the decimal values than are necessary for the computation.

Example 5. 9.435 V, when rounded off to the nearest 1/100 V, equals 9.44 V. Since the third decimal digit is followed by at least one digit other than zero and the value of this digit is more than 5, the last digit to be retained (hundreds) is increased by 1.

If the original value is changed to 9.435 0 V and is rounded off to two decimal places, its value is 9.43 V. In this case, the value of the last two digits is not more than 5.

CONVERSION FACTORS IN METRICATION

Tables of conversion factors are necessary and helpful when converting values between different measurement systems. Computations using these conversion factors will be required until all nations use a single acceptable system of measurement. A number of selected tables are included in this unit. These tables help to convert measurements from the customary system to SI metrics and vice versa.

The conversion factor tables contain up to eight place decimal values. As stated previously, the number of places used and the rounding off process for decimal values are determined by the degree of accuracy required.

The meter is usually given as the standard of length. Smaller and larger units of measure may be used by reviewing the relationships of values in the decimal system and the value assigned to the prefixes.

The reverse process of conversion may be performed by using the reciprocal of any multiplier as the divisor. For example, to convert from centimeters to inches, multiply the number of centimeters by 0.4 (rounded to one decimal place). Thus, 2.5 cm (approximately) = 1 inch. If 2.5 is used as the reciprocal, then 1 cm = 1/2.5 inch or 0.4 inch (approximately).

Conversion Processes and Conversion Factors

Conversions may be made from decimal multiple and submultiple values of SI units. The decimal point is moved according to the prefix of the unit. For example, for the base unit of length (the meter), submultiple decimal units such as milli (0.001) and centi (0.01), and multiple units such as kilo (1000), may be used to solve common problems. Thus, in place of the value 1 m = 39.37 in, the conversion factor in a table may be changed to 1 mm = 0.039 37 inch, 1 cm = 0.393 7 inch, or 1 km = 39 370 inches. The way the conversion factor is expressed depends on the prefix used when a numerical value is stated.

Two simplified conversion tables, tables 7-6 and 7-7, illustrate the type of information that is available. Each table shows metric and customary units and values that are used in everyday problems involving measurements of length and mass. Additional

Metric Units		
Conversion From → To		Conversion Factor (Multiply by)
meters (m)	inches	39.370 08
	feet	3.280 840
	yards	1.093 613
	miles	0.000 621 37

Customary Units		
Conversion From → To		Conversion Factor (Multiply by)
miles	inches	63 360
	centimeters	160 934.4
	meters	1 609.344
yards	meters	0.914 4
	kilometers	0.000 914 4
feet	meters	0.304 8
	kilometers	0.000 304 8
inches	centimeters	2.54

Table 7-6 Units of length

52 Systems of Measurement in Science

Metric Units		
Conversion		Conversion Factor (Multiply by)
From →	To	
kilograms (kg) →	grains	15 432.36
	avoirdupois drams	564.383 4
	avoirdupois ounces	35.273 96
	avoirdupois pounds	2.204 623
	Troy ounces	32.150 75
	Troy pounds	2.679 229

Customary Units		
Conversion		Conversion Factor (Multiply by)
From →	To	
avoirdupois pounds (lb) →	kilograms	0.453 592 37
	metric tons	0.000 453 59
	grains	7 000
	avoirdupois drams	256
	avoirdupois ounces	16
	Troy ounces	14.583 33
	Troy pounds	1.215 28

Table 7-7 Units of mass

conversion tables for other measurements such as area, capacity, or volume appear in the Appendix. Still other tables are available from professional engineering and scientific societies, handbooks, industrial literature, governmental agencies involved in weights and measures, and private technical publications.

Another form of table gives conversion factors for measurements arranged according to major categories. Table 7-8 is an abridged table which shows the multiplication factors by which a customary measurement may be either changed or converted to an equivalent metric unit.

Metric Conversion Factors for Customary Measurement Units

Category	Conversion		Multiply by
	From	To	
acceleration	ft/s^2	m/s^2	0.304 8*
	in/s^2	m/s^2	$2.540\ 0 \times 10^{-2}$*
area	ft^2	m^2	$9.290\ 3 \times 10^{-2}$
	in^2	m^2	$6.451\ 6 \times 10^{-4}$*
density	g/cm^3	kg/m^3	$1.000\ 0 \times 10^{3}$*
	lb (mass)/ft^3	kg/m^3	16.018 5
	lb (mass)/in^3	kg/m^3	$2.768\ 0 \times 10^{4}$

*Exact

Table 7-8 Metric conversion factors for customary measurement units

Metric Conversion Factors for Customary Measurement Units

Category	Conversion From	To	Multiply by
energy	Btu (thermochemical)	J	$1.054\ 3 \times 10^3$
	cal (thermochemical)	J	$4.184\ 0*$
	eV	J	$1.602\ 1 \times 10^{-19}$
	erg	J	$1.000\ 0 \times 10^{-7}*$
	ft · lb (force)	J	$1.355\ 8$
	kWh	J	$3.600\ 0 \times 10^6 *$
	Wh	J	$3.600\ 0 \times 10^3 *$
flow, liquid and solid	ft^3/min	m^3/s	$4.719\ 5 \times 10^{-4}$
	ft^3/s	m^3/s	$2.831\ 7 \times 10^{-2}$
	in^3/min	m^3/s	$2.731\ 2 \times 10^{-7}$
	lb (mass)/s	kg/s	$0.453\ 6$
	lb (mass)/min	kg/s	$7.559\ 9 \times 10^{-3}$
	tons (short, mass)/h	kg/s	$0.252\ 0$
force	dyne	N	$1.000\ 0 \times 10^{-5}*$
	kg (force)	N	$9.806\ 6$
	lb (force)	N	$4.448\ 2$
heat	Btu (thermochemical)/ft^2	J/m^2	$1.134\ 9 \times 10^4$
	cal (thermochemical)/cm^2	J/m^2	$4.184\ 0 \times 10^4 *$
	ft^2/h	m^2/s	$2.580\ 6 \times 10^{-5}$
length	ft	m	$0.304\ 8*$
	in	m	$2.540\ 0 \times 10^{-2}*$
	μ (micron)	m	$1.000\ 0 \times 10^{-6}*$
	mil	m	$2.540\ 0 \times 10^{-5}*$
mass	lb (mass, avoirdupois)	kg	$0.453\ 6$
	oz (mass, avoirdupois)	kg	$2.835\ 0 \times 10^{-2}$
	ton, long = 2 240 lb (mass)	kg	$1.016\ 0 \times 10^3$
	ton, metric	kg	$1.000\ 0 \times 10^3 *$
	ton, short = 2 000 lb (mass)	kg	$0.907\ 2 \times 10^3$

*Exact

Table 7-8 Metric conversion factors for customary measurement units (continued)

Metric Conversion Factors for Customary Measurement Units

Category	Conversion From	Conversion To	Multiply by
power	Btu (thermochemical)/min	W	17.572 5
	cal (thermochemical)/min	W	$6.973\ 3 \times 10^{-2}$
	erg/s	W	$1.000\ 0 \times 10^{-7}$*
	ft · lb (force)/min	W	$2.259\ 7 \times 10^{-2}$
	hp (550 ft · lb/s)	W	$7.457\ 0 \times 10^{2}$
pressure (stress)	atm (760 torr)	N/m²	$1.013\ 2 \times 10^{5}$
	dyne/cm²	N/m²	0.100 0*
	g (force)/cm²	N/m²	98.066 6*
	kg (force)/cm²	N/m²	$9.806\ 6 \times 10^{4}$
	lb (force)/in² (or psi)	N/m²	$6.894\ 8 \times 10^{3}$
	lb (force)/in² (or psi)	kg (force)/mm²	$7.030\ 7 \times 10^{-4}$
	torr (mmHg at 0 degrees C)	N/m²	$1.333\ 2 \times 10^{2}$
velocity	ft/min	m/s	$5.080\ 0 \times 10^{-3}$*
	in/s	m/s	$2.540\ 0 \times 10^{-2}$*
	mph	m/s	0.447 0
	mph	km/h	1.609 3
volume	ft³	m³	$2.831\ 7 \times 10^{-2}$
	in³	m³	$1.638\ 7 \times 10^{-5}$
	liter	m³	$1.000\ 0 \times 10^{-3}$*
temperature	degree C	K	$t_K = t_C + 273.15$
	degree F	K	$t_K = (t_F + 459.67)/1.8$

*Exact

Table 7-8 Metric conversion factors for customary measurement units (continued)

SUMMARY

- The American National Standards Institute (ANSI) is an advisory organization to industry. ANSI also represents the United States in dealing with metrication issues within the International Standards Organization.
- Metrication relates to any program or process of converting to the International System of Units (SI).

- The base ten notation system is used in the International Metric System (SI). Multiples (+) and submultiples (−) of the powers of ten simplify the mathematical processes involved in solving scientific problems and simplify the statement of values having large numbers of digits.
- Rules adopted for writing and expressing values and processes in SI cover the base, supplementary, and derived units of measure. Some of the major rules are:
  — insignificant digits are eliminated.
  — prefixes express the order of magnitude of a value.
  — symbols derived from proper names and the numerical prefixes of giga (G), tera (T), and mega (M) are capitalized.
  — number values are placed in groups of three digits and are written in this manner, without commas.
  — values may be rounded off by increasing the last required digit by 1 if the value of the digits beyond this position is more than 5.
- SI metrics has the following advantages:
  — Only one unit is used for each physical quantity.
  — A standardized and defined set of symbols and abbreviations means that uniform communication and interpretation are possible.
  — It is a simplified system of computations involving powers of ten, multiples and submultiples of the base units, prefixes, and conversion factors.
  — It is a coherent system which yields accurate results.
  — Essential derived units make the system flexible to meet consumer and producer needs.
- The seven base units in SI metrics are as follows:
  — meter (m), the unit of length,
  — kilogram (kg), the unit of mass,
  — second (s), the unit of time,
  — ampere (A), the unit of electrical current,
  — degree kelvin (K), the unit of thermodynamic temperature,
  — candela (cd), the unit of luminous intensity,
  — mole (mol), the unit of substance of a system.
- Plane and solid angles are represented in SI metrics by two supplementary units: the radian (rad) and the steradian (sr).
- Derived units represent other acceptable standards of measurement. These additional units are derived from the base and supplementary units.
  — For example, the derived units of force, work, power, and pressure relate to the base unit of mass; the derived units of potential, resistance, capacitance, inductance, quantity, and magnetic flux relate to the base electrical unit, the ampere. Derived units are applied daily in practical applications.

56 Systems of Measurement in Science

- SI metrics requires a working knowledge of the powers of ten (scientific notation system); prefixes and values; base, supplementary, and derived units of measurement; the expression of quantities and values in terms of technical standards for usage; and the application of conversion factors.
  — Tables of conversion factors list the numerical values to be used to convert a unit of measurement from a customary unit to SI metrics, and vice versa.
  — The number of decimal digits to be used in computations and for rounding off values is determined by the required degree of accuracy.

ASSIGNMENT UNIT 7 METRICATION: SI METRIC UNITS OF MEASURE

■ ORGANIZATION AND FUNCTIONS OF THE GOVERNMENT AND THE PRIVATE SECTOR IN METRICATION

1. State briefly how the United States Metric Board relates to metrication in this nation.
2. Identify three major functions of the United States Metric Board.

■ BASE TEN SCIENTIFIC NOTATION SYSTEM

1. Four prefixes that are used with SI metric units and positive factors and four symbols used with negative factors are given in the following table.
   a. Give the appropriate symbol, the base ten exponent, and the multiplication factor for items A through D.
   b. Complete the table for items E through H by giving the prefix, base ten exponent, and multiplication factor.

	Prefix	Symbol	Base Ten Exponent	Multiplication Factor
A	giga			
B	mega			
C	kilo			
D	hecto			
E		d		
F		m		
G		μ		
H		p		

Metrication: SI Metric Units of Measure

2. Use the measurement units and exponential factors given in the table for problems A through F.
   a. Solve each problem. Give the computed numerical value and base ten factor.
   b. Restate the answer in terms of its numerical value.

Problem	Quantity	Solution	
		Quantity and Base Ten Exponent	Numerical Value
A	5.4 g (10^2) + 6.2 g (10^2)		
B	5.4 m (10^2) × 6.2 m (10^2)		
C	(5.4 cm^2 · 10^4) · (0.62 cm · 10^{-2})		
D	(5.4 MV · 10^{-4}) · (6.2 MW · 10^2)		
E	$\dfrac{5.4 \text{ kg} \cdot 10^6}{2.7 \text{ kg} \cdot 10^4}$		
F	$\dfrac{5.4 \text{ Hz} \cdot 10^6}{2.7 \text{ Hz} \cdot 10^{-2}}$		

Guide for Style and Usage: SI Metric System

1. Complete the following table as indicated.
   a. Write each computed value (A through E) according to the rules for SI metric measurements.
   b. Round off each of the values (A-E) to the degree of accuracy indicated in each case. State each rounded off value as an SI metric number.

   NOTE: The answer is given for Part A to show what is required.

	Computed Value	Stated as SI Metric Value	Accuracy (Decimal Places)	SI Metric Numerical Value
A	1004.25	1 004.25	1	1 004.2
B	100004.2543		2	
C	1004.249655		1	
D	1000042.2512722		4	
E	100004.25672		2	

■ **ADVANTAGES OF SI METRICS**

1. List three distinct advantages of SI metrics over other metric systems or the British measurement system.
2. Briefly state one economic advantage and one scientific advantage of the process of metrication in the United States.

58 Systems of Measurement in Science

■ **BASE, SUPPLEMENTARY, AND DERIVED SI METRIC UNITS**

1. Complete the following table for the seven base SI metric units.

   a. List the seven basic quantities of SI metrics.
   b. Identify the SI base unit for each quantity.
   c. Give the symbol for each SI base unit.

	Quantity (a)	Base Unit (b)	Symbol (c)
1			
2			
3			
4			
5			
6			
7			

2. State briefly the differences between:

   a. Base units and supplementary units in SI metrics.
   b. Derived units and base units in SI metrics.

3. a. Select and name one base unit.
   b. List three quantities that must be measured that relate to the base unit.
   c. State the derived unit of measurement for each of the three quantities.
   d. Give the formula for finding the value of each derived unit.

■ **CONVERSION FACTORS AND PROCESSES**

Refer to a Table of Conversion Factors for both SI metric and customary units for the quantities of categories A through F in the following table. For problems 1 through 23, either the SI metric unit values or the customary unit values are given.

1. Indicate the conversion for each problem.

2. Compute the value of each measurement and round off each answer to the number of significant places indicated. Record and properly designate the unit in which the answer is given.

Metrication: SI Metric Units of Measure

Category			Measurement		Conversion Factor (1)	Computed Measurement	
			Given Unit Value	Required Unit Value		Significant Places	Rounded Off Value and Unit Designation (2)
A	Length	1	5 in	millimeters		2	
		2	27 yd	meters		3	
		3	16 ft 3 in (16 1/4′)	meters		1	
		4	184 km	miles (statute)		2	
		5	232.5 mm	inches		3	
B	Mass	6	100 kg	pounds/mass (avoirdupois)		2	
		7	100 kg	pounds/mass (apothecary-Troy)		3	
		8	20 000 grains	kilograms		3	
		9	62.4 lb · m (Troy)	kilograms		2	
	Mass Area	10	17 400 kg/m^2	ounce · mass/yard^2		2	
	Mass Time	11	12.66 kg/s	pounds · mass/minute		2	
	Mass Volume	12	24 lb · m/ft^3	kilogram/meter^3		3	
		13	2 156.9 kg/m^3	pound · mass/gallon (US)		1	
C	Pressure	14	20 psi	pascal		1	
		15	32 Pa	kilogram (force)/meter^2		3	
D	Power	16	100 Btu/s	kilowatts		2	
		17	20 000 watt	horsepower (metric)		1	
		18	118 kc/min	watts		2	
E	Temperature	19	110 degrees C	degree Kelvin		2	
		20	246.15 degrees K	degree Celsius		2	
		21	104 degrees F	degree Celsius		nearest whole number	
F	Velocity	22	80 km/h	miles per hour (mph)		nearest whole number	
		23	80 mi/h	kilometer/hour		nearest whole number	

Achievement Review of Systems of Measurement

SYSTEMS OF MEASUREMENT IN SCIENCE

■ UNITED STATES CUSTOMARY, BRITISH AND SI METRIC MEASUREMENT SYSTEMS

Complete statements 1 to 10 by selecting the correct word(s) or value.

1. The (yardstick) (steel rule) (meter stick) is the most accurate device for taking linear measurements to an accuracy of 1/64 or 1/100 of an inch.
2. Measurements to accuracies of 0.0001 inch may be made with a (steel tape) (meter stick) (micrometer).
3. Precision laboratory measurements may be made within practical working limits of (0.1 in) (0.000 002 in) (0.000 000 002 in).
4. The system of measurement commonly used by industry and technologists worldwide is the (English) (SI metric) system.
5. The square inch is (larger than) (smaller than) (the same size as) a square foot.
6. A cubic meter is (larger than) (smaller than) (the same size as) a cubic foot.
7. A liter is (larger than) (smaller than) (the same size as) a quart.
8. A cubic yard is (larger than) (smaller than) (the same size as) a cubic meter.
9. The volume of irregularly shaped objects is usually (computed) (located in tables) (found by the displacement method).
10. One thousand grams equal (one metric ton) (2.2 pounds, approximately) (20 centigrams).
11. Compute the cubical volume of cubes A and B, rectangular containers C and D, and cylinders E and F (English measure). Use $\pi = 22/7$.

	A	B	C	D		E	F
Length	5″	1′-6″	10″	1 2/3 yd	Diameter		3′-6″
Depth	5″	1′-6″	5″	2′-4″	Radius	7″	
Height	5″	1′-6″	2 1/2″	1′	Length	10″	36 yd

12. Compute the volumes (in the metric system) of rectangular solids A and B, and cylindrical solids C and D. Round off each answer to the nearest whole number.

	A	B		C	D
Length	2 m	8 m	Diameter	14 mm	
Depth	4.5 m	10 dm 50 cm (10.5 dm)	Radius		1 m 5 dm (1.5 m)
Height	6.2 m	1 m 20 cm (1.02 m)	Length	25 mm	2 m 6 dm 6 cm (2.66 m)

13. Determine the volume in liters and the weight of water in kilograms required to fill containers A, B, and C to the height indicated. Use $\pi = 22/7$ and round off answers to the nearest whole number.

	Shape of Container	Inside Dimensions	Liquid Height
A	Square	10 cm	5 cm
B	Rectangular	10 m × 5 dm	2 dm 5 cm (2.5 dm)
C	Round	2 m 8 dm (2.8 m) diameter	1 m

■ **METRICATION: SI METRIC UNITS OF MEASURE**

1. State two advantages and two disadvantages to industries resulting from the conversion of products and processes from the United States Customary and British system to the SI metric system of measurement.

2. For quantities A–D in the following table:
   a. State each quantity in terms of its base ten value.
   b. Compute the numerical value (expressed in terms of the quantity and the exponent).
   c. Restate the answer as a numerical value.

	Quantity and Process	Base Ten Value (a)	Quantities and Exponent (b)	Numerical Value (c)
A	20 000 Hz · 17 900 Hz			
B	108 000 kV · 12 600 kV			
C	$\dfrac{10\ 800\ \text{kg}}{27\ \text{kg}}$			
D	$\dfrac{2\ 160\ \text{kg}}{2.7(10^{-2})\ \text{kg}}$			

Systems of Measurement in Science

■ INTERNATIONAL SYSTEM OF UNITS (SI METRICS)

Secure a table of conversion factors for both SI metric units and customary units of measurement for length, mass, pressure, power, temperature, and velocity.

1. Insert the conversion factors required to change from the given value to the required unit value for problems 1-9 in the following table.
2. Compute the value of each measurement and round off the value to the stated number of significant places. Record the value and unit designation in the following table.

	Measurement Category		Given Value	Required Unit Value	Conversion Factor	Required Significant Places	Rounded-off Value and Units
A	Length	1	32 ft 6 in (32 1/2')	meters		2	
		2	9 224 km	miles (statute)		whole number	
B	Mass	3	220 kg	pounds/mass (avoirdupois)		2	
		4	486 lb (mass) (Troy)	kilograms		2	
	Mass Time	5	18.99 kg/s	lb mass/min		3	
C	Pressure	6	56 Pa	kg (force)/m^2		2	
D	Power	7	177 kc/min	watts		1	
E	Temperature	8	216.72°K	°Celsius		2	
F	Velocity	9	62 km/h	m/s		1	

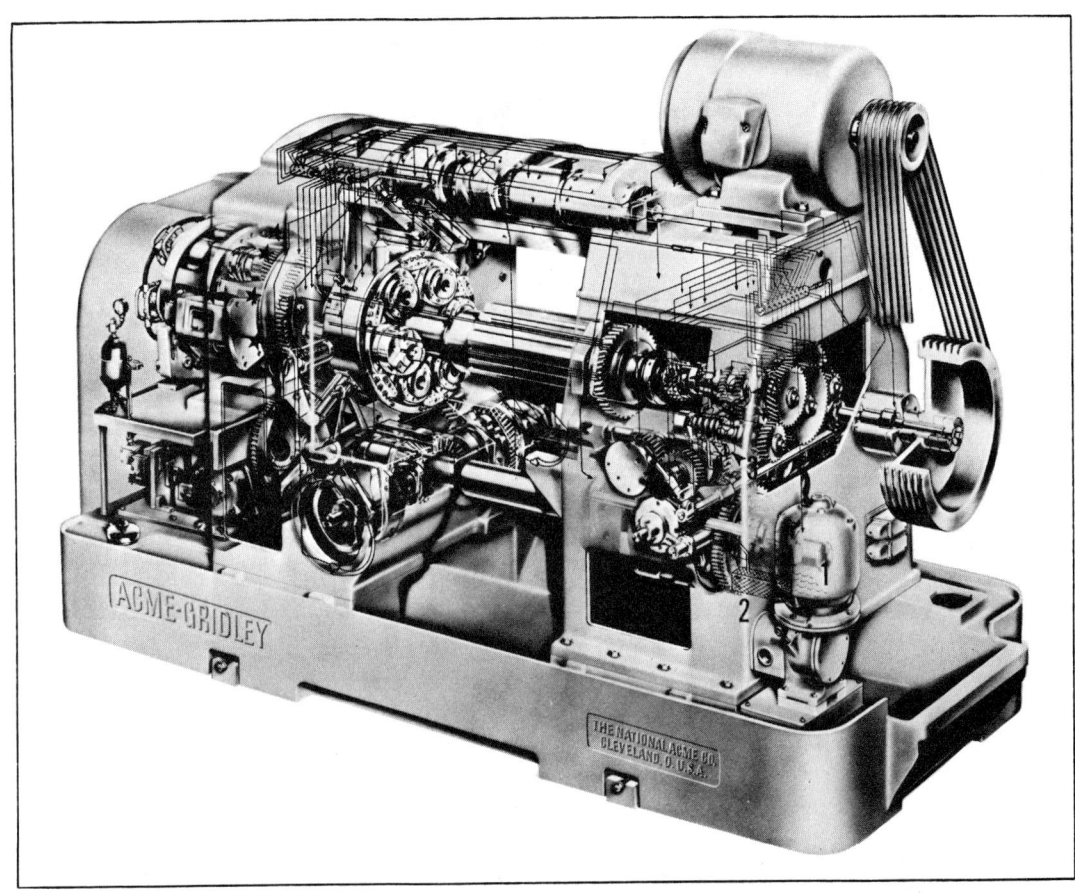

Major components of a modern multiple spindle production turning machine involve the principles of forces, balance and equilibrium, motion, simple and compound machines, pneumatics, and fluid power (Courtesy National Acme Company, Cleveland, Ohio)

Section 3
Mechanics, Machines, and Wave Motion

Unit 8 Forces and Their Effects

A study of any machine or mechanism shows that each is made up of a number of movable parts. These parts transform a given motion to a desired motion. In other words, these machines perform work. *Work* is done when motion results from the application of force. Thus, a study of mechanics, machines, and wave motion deals with forces and the effects of forces on bodies.

A *force* is a push or pull. The effect of a force either changes the shape or motion of a body or prevents other forces from making such changes. Every force produces a stress in the part on which it is applied. Forces may be produced by an individual using muscular action or by machines with mechanical motion.

For example, when a person cuts a board, a downward and a forward push is applied. It is this muscular force which causes the saw teeth to cut. As another example, the jaws of a drill chuck may be tightened with a chuck wrench by applying a force by hand.

■ KINDS OF FORCE: TENSION, COMPRESSION, TORQUE, AND SHEAR

Forces are produced by physical or chemical change, gravity, or changes in motion. When a force is applied which tends to stretch an object, it is called a *tensile force*. A part experiencing a tensile force is said to be in *tension*, figure 8-1.

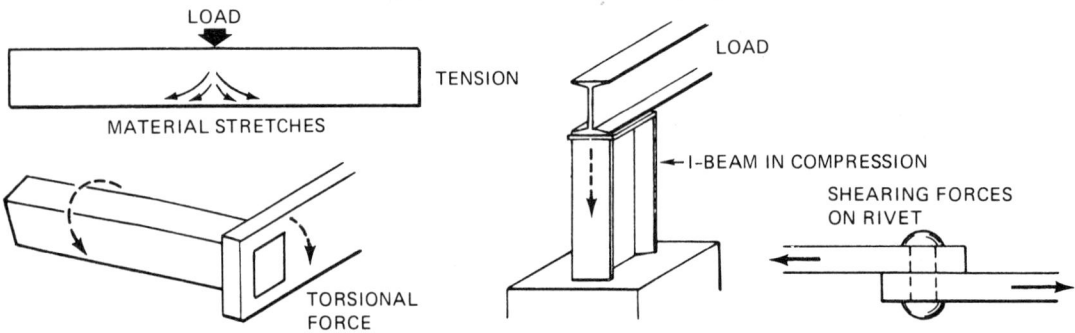

Fig. 8-1

A force can also be applied which tends to shorten or squeeze the object. Such a force is a *compressive force*. A part experiencing a compressive force is in *compression*, figure 8-1.

A third force is known as a *torsional force*, or a *torque* since it tends to twist an object. Still another kind of force, which seems to make the layers or molecules of a material slide or slip on one another, is a *shearing* force.

Each of these forces may act independently or in combination. For example, a downward force applied on a vertical steel beam tends to compress the beam. If this beam is placed in a horizontal position and a load is applied in the middle, the bottom of the beam tends to stretch and is in tension. At the same time, the top area is being pushed together in compression. If the compressive and tensile forces are great enough to make the layers of the material slide upon each other, a shearing force results.

The turning of a part in a metal lathe is another example of several forces in action, figure 8-2. As the work revolves and the cutting tool moves into the work, the wedging action of the cutting edge produces a shear force. This force causes the metal to seem to flow off the work in the form of chips. If this workpiece is held between the centers of the lathe, the centers exert a compressive force against the work. The lathe dog which drives the work tends to produce a shearing force. The pressure of the cutting tool against the work produces tension and compression, as well as a shearing action.

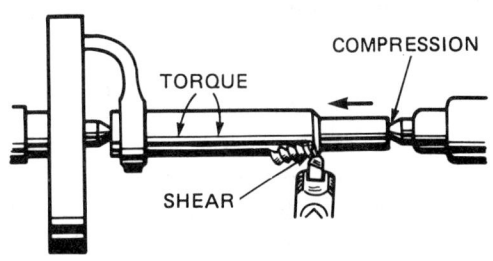

Fig. 8-2

■ EFFORT FORCES AND RESISTANCE FORCES

An *effort force* refers to the force applied to a part. The resistance of the part to this applied force is the *resistance force*. For a 100% efficient machine, it can be shown that the effort force multiplied by the distance through which it acts is equal to the resistance force multiplied by its distance, at equilibrium, figure 8-3.

This relationship is the basis for comparing effort (E), effort distance (ED), resistance (R), and resistance distance (RD). Such a comparison can be stated mathematically as an equation.

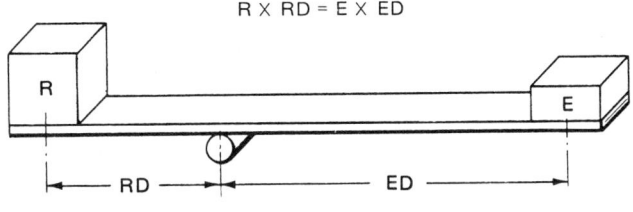

Fig. 8-3

66 Mechanics, Machines, and Wave Motion

Effort (E) × Effort Distance (ED) = Resistance (R) × Resistance Distance (RD)

Thus, (E) × (ED) = (R) × (RD)

or, $\dfrac{E}{R} = \dfrac{RD}{ED}$

The effort torque is E × ED and the resistance torque is R × RD. It can be seen that *torque* is the product of a force times the distance through which the force acts. When the effort is expressed in pounds of pull, and the distance through which it acts is given in inches, then the effort-torque is in pound-inches (the product of pounds × inches). Note that the effort and resistance forces are measured in like units (pounds); similarly, the distances are also measured in like units (inches). The distances may be given in feet as well.

The formula E × ED = R × RD can be used to find any one missing value when the three remaining values are given. For example, if a force of 25 pounds (E) is applied to an open-end wrench and the effort force operates through a distance of four inches (ED), the resistance offered by a square metal plug moving a distance of one inch is:

E × ED = R × RD

25 pounds × 4″ = R × 1″

R = 100 pounds (45.4 kg)

The resistance offered by the plug to the force is 100 pounds (45.4 kg). The resistance and effort forces can be compared mathematically because they are in the same unit of measurement. For the same reason, the resistance and effort distances may be compared as a ratio.

■ MECHANICAL ADVANTAGE OF FORCE AND SPEED

Mechanical Advantage of Force

Whenever a small force is used to move a larger force, there is a gain which is called *mechanical advantage*. In the previous example, an effort force of 25 pounds (11.4 kg) moves a resistance force of 100 pounds (45.4 kg). The mechanical advantage of force (MA_f) is the ratio of the resistance force to the effort force.

$$MA_f = \dfrac{R}{E}$$

Substituting values from the example yields:

$$MA_f = \dfrac{100}{25} = 4 \text{ or, } \dfrac{45.4}{11.4} = 4$$

Thus, the mechanical advantage of force is 4.

Mechanical Advantage of Speed

The mechanical advantage of speed or distance (MA_s) is the ratio of the resistance distance to the effort distance.

$$MA_s = \frac{RD}{ED}$$

For the previous example, ED = 4″ (10.16 cm) and RD = 1″ (2.54 cm).

$$\text{Therefore,} \frac{RD}{ED} = \frac{1''}{4''} \text{ or } \frac{1}{4} \text{ or, } \frac{2.54 \text{ cm}}{10.16 \text{ cm}} = \frac{1}{4}$$

The mechanical advantage in speed is 1/4. It is obvious that the mechanical advantage in force is compensated for by the loss in speed.

The principles underlying the nature and kinds of forces and the mechanical advantages of distance and force are basic to an understanding of the units which follow.

■ WORK, POWER, AND EFFICIENCY

Work is defined as the product of a force and the distance through which the force acts. Work is expressed in terms of linear units and weight units, such as inch-pounds or foot-pounds. Joules (J) are used in metric. In general, foot-pounds are converted to joules by rounding off the factors of 0.738 to 0.74 and 1.3558 to 1.4, respectively, and then multiplying.

If a wood plank weighing 60 pounds is lifted 4 feet from the floor, the work done equals 60 pounds × 4 feet or 240 foot-pounds (336 joules), figure 8-4. Work is performed in lifting the plank because a motion results from the application of a force. However, if the plank weighs 600 pounds and an hour is spent trying to lift it without success, then, by definition no work is done.

The distance used to compute work is the distance the object moves in the same direction as the applied force. The distance used in torque calculations is perpendicular to the direction of the force.

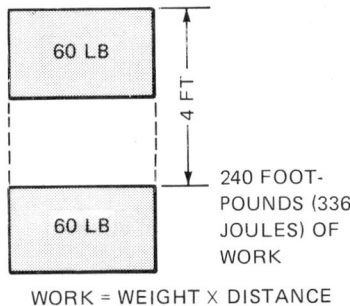

WORK = WEIGHT × DISTANCE

Fig. 8-4

Power

The term *power* is applied to the rate at which work is done. In other words, a third unit of measurement (time) is applied to work. If a lift truck raises a 6600-pound casting 5 feet in one minute, the rate of doing work is 33 000 foot-pounds per minute; the amount of work done is 33 000 foot-pounds.

The rate at which work may be done is compared to a universally accepted standard. The *horsepower* (hp) is the standard unit of measurement used to express 33 000 foot-pounds of work per minute or 550 foot-pounds of work per second, figure 8-5. The power of motors and engines in horsepower units can be determined by dividing the known or computed rate in foot-pounds per minute by 33 000.

$$\text{hp} = \frac{\text{work (foot-pounds)}}{33\,000 \times \text{time (minutes)}} \text{ or, } 1 \text{ hp} = 746 \text{ watts (W)}$$

68 Mechanics, Machines, and Wave Motion

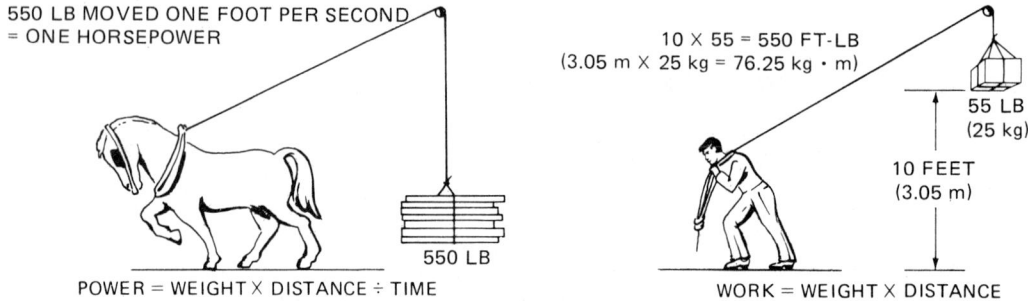

Fig. 8-5

Efficiency

A machine transforms power from one form to another. There is some loss in this process as the usable power output is always less than the power input. The ratio of the output to the input power (expressed in the same units of measure) is known as the *efficiency* of the machine. Efficiency is given as a percent in many instances, and is always less than 100% (or 1.00). Friction and heat losses account for most of the losses in efficiency.

A crane lifts a 500-pound girder 5 feet. The work done is 2500 foot-pounds (3500 joules). To do this work, a cable moves through a distance of 20 feet and exerts a force of 150 pounds. The input is 20 × 150 or 3000 foot-pounds (4200 joules); the output is 5 × 500 or 2500 foot-pounds (3500 joules). The efficiency can be determined as follows:

$$\text{Efficiency} = \frac{\text{Output}}{\text{Input}} = \frac{2500}{3000} = 83\ 1/3\%\ \text{or,}\ \frac{3500}{4200} = 83\ 1/3\%$$

For this problem, the mechanical advantage of force is 500/150 or 3 1/3. The mechanical advantage of speed is 5/20 or 1/4. It is possible to lift heavy weights using lighter forces when there is a loss in the distance or speed.

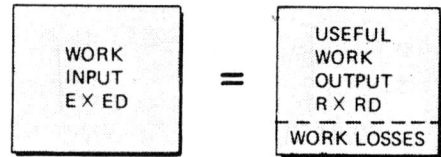

Fig. 8-6

SUMMARY

- A force either changes the shape or motion of a body or prevents other forces from making such changes.
- A force may be produced by physical or chemical change, gravity, or change in motion.
- The four kinds of forces most used in industry are: (1) tensile, (2) compressive, (3) torsional and (4) shearing.

- The product of the effort force and the distance through which it acts is equal to the product of the resistance force and its distance.
- The mechanical advantage of force is the ratio of the resistance (weight to be lifted) to the effort (force needed for the job), expressed in the same units of measure.
- The mechanical advantage of speed is the ratio of the resistance distance to the effort distance.
- Work is the product of a force and the distance through which it acts, expressed in linear and weight units.
- Power is the rate of doing work. The standard horsepower is the rate of doing 33 000 foot-pounds of work a minute, or 746 watts.
- Efficiency is the ratio (percent) of output to input.

ASSIGNMENT UNIT 8 FORCES AND THEIR EFFECTS

■ PRACTICAL PROBLEMS ON FORCES, WORK, POWER, EFFICIENCY

Forces and Their Effects

1. State the kind of force caused by each of the following:

   a. filing on a lathe
   b. cutting lips of a drill
   c. jaws of a vise
   d. plunger on hydraulic press

2. The cutting edge of a lathe tool is positioned 30° toward the chuck. The end of the work is supported by a tailstock center. Make a simple sketch and identify with arrows the kind of force exerted by the lathe chuck, center, and cutting edge of the tool as the cut is taken.

Effort Forces, Resistance Forces, and Mechanical Advantage

For statements 1 to 6, determine which are true (T) and which are false (F).

1. An effort force refers to the resistance a part offers.
2. Effort and resistance forces are given in the same units of measurement.
3. When E = 10 lb, ED = 10 in, and R = 20 lb, then RD = 15 in.
4. The mechanical advantage of force of 100 newtons (N) moving a 200-newton (N) part is 2.
5. The mechanical advantage of speed in moving an object 2 meters with an effort distance of 4 meters is 1/2.
6. The mechanical advantage is equal to the resistance divided by the resistance arm.

Mechanics, Machines, and Wave Motion

Work, Power, and Efficiency

Indicate the letter representing the words, phrases, or formulas which best complete statements 1 to 7.

1. Work is done when
    a. a force is exerted.
    b. motion is produced.
    c. effort is applied.

2. Work is equal to
    a. output ÷ input.
    b. resistance ÷ effort.
    c. force × distance.

3. Work input equals
    a. effort × resistance.
    b. resistance ÷ effort.
    c. effort × effort distance.

4. Efficiency equals
    a. output ÷ input.
    b. force × distance.
    c. input ÷ output.

5. Heat and friction losses cause
    a. a decrease in efficiency.
    b. an increase in efficiency.
    c. no change in efficiency.

6. Horsepower is a measure of
    a. foot-pounds of work.
    b. work ÷ time.
    c. efficiency × time.

7. One horsepower equals
    a. 550 foot-pounds per second.
    b. 100 kW.
    c. 66 000 joules per second. Note: J = 0.738 ft-lb.

8. Compute the horsepower required to deliver
    a. 66 000 foot-pounds per minute.
    b. 2750 foot-pounds per second.
    c. 100 kW.
    d. 66,000 joules per second.

9. Which letters indicate a situation where work is being performed?
   a. Measuring work length.
   b. Chuck jaws holding work.
   c. Lifting a saw horse.
   d. Tightening a motor nut.
   e. Bolt holding parts together.
   f. Computing wire size.

10. Calculate the work done by a 320-N (newton) force which moves a machine component a distance of 10 m (meters).

11. Convert the work done by the 320-N force and the 10-m distance to its equivalent value in customary foot-pound units. Note: Use 1 J = 0.74 ft-lb.

Unit 9 Balance and Equilibrium — Parallel and Angular Forces

A mechanic must understand the conditions which affect the balance and stability of moving or stationary parts so as to determine when the work is properly anchored. If an object is to be balanced, it must be in *equilibrium;* that is, there must be no unbalanced forces tending to produce motion. An object balanced while at rest is said to be in *static balance;* an object balanced while in motion is in *dynamic balance.*

Materials which are not in static balance tend to turn or rotate in an effort to achieve balance. Forces which produce rotation are called *torque* (torsional) *forces* or *moment forces.* A moment is the product of the force and the distance through which it acts. This distance is the length of a perpendicular drawn from the axis of rotation to a line representing the direction in which the force is applied. The perpendicular passes through the point of application of the force. Since force is generally measured in pounds and distance is measured in feet or inches, the moment is expressed in foot-pounds (often referred to as pound-feet). Moments may produce clockwise or counterclockwise rotation. A body is in equilibrium when the clockwise and counterclockwise forces are equal and the body is free to rotate.

■ DYNAMIC AND STATIC BALANCE AND EQUILIBRIUM

It is much easier to balance a part statically than it is to balance it dynamically. For example, a two-pound weight is placed at the right end of a short beam, figure 9-1. If the weight is placed three feet from the point at which the beam is held, the beam has a tendency to turn clockwise. The moment is equal to the force (two pounds) × the distance (three feet) or six pound-feet.

If another weight of three pounds is placed two feet to the left of the point at which the beam is pivoted, the beam will turn in a counterclockwise direction. In this case, the moment is equal to the force (three pounds) × the distance (two feet) or six foot-pounds.

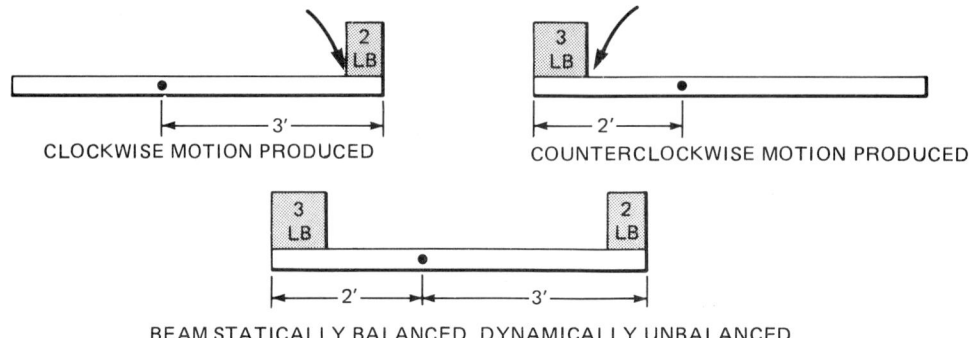

Fig. 9-1

This beam is in static balance since the moment which tends to turn the body in a clockwise direction (six pound-feet) is equal to the moment tending to rotate the body counterclockwise (6 pound-feet). However, there will be considerable vibration if the whole system turns. Because the system is out of balance dynamically, there will be undue wear on the moving parts, excessive vibration, and uneven motion. For these reasons, it is important that rotating parts be in dynamic balance. The location of the points where a part is out of balance may be determined by computation, by trial and error, or with special balancing equipment.

The degree to which a moving part is balanced depends upon its use. A large turbine rotor turning at high speeds requires precision balancing to a far greater degree than does the wheel of an automobile. Dynamic balancing may be done either by removing material from the heavy side of a device or by attaching weights to the light side, as in the case of automobile wheels.

Center of Gravity

The technician and engineer work with many irregularly shaped parts. It is important to determine the point where each of these parts is statically balanced, regardless of the position in which it is placed. This point is known as the *center of gravity*.

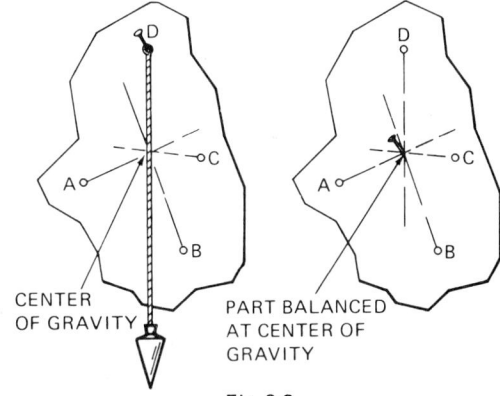

Fig. 9-2

If it is necessary to find the center of gravity of an irregularly shaped piece of cardboard, the procedure is very simple, figure 9-2.

1. Punch four or five holes near the outer edges of the cardboard.
2. Place the cardboard on a nail and drop a plumb line from this hole.
3. Mark the line of the plumb line on the cardboard.
4. Repeat this process from each of the other holes.
5. The point at which the plumb lines intersect is the center of gravity. This point may be tested by inserting a pin through it and testing the cardboard for balance.

The principle governing the location of the center of gravity is important to designers and is used daily in the loading and balancing of irregularly shaped parts. If the cardboard in the previous example has considerable thickness and another material is substituted for it, the center of gravity will be located between the two outer surfaces at the intersecting point found with the plumb line.

Stability

An object may be in *stable* or *unstable equilibrium* depending upon the location of the center of gravity. Stability refers to the condition of balance of the object. When

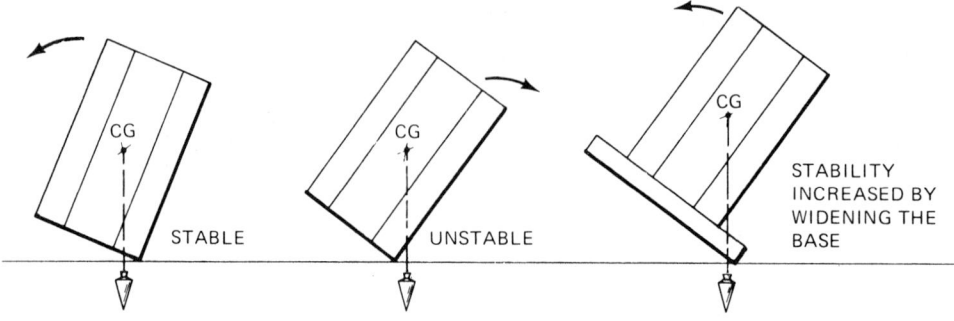

Fig. 9-3

an object is tipped and it returns to its original position when released, it is in stable equilibrium. For any object in stable equilibrium, a plumb line dropped from the center of gravity will fall within the base of the object, figure 9-3.

The object is in unstable equilibrium when it is tipped so far that a plumb line from the center of gravity falls outside the base of the object. The center of gravity of an object seeks the lowest position possible. When a stable object is tipped, its center of gravity is raised. Thus, the object tends to return to its original position. If tipping the object lowers its center of gravity, it will not return to its original position. The center of gravity must be raised if the object is to return to its original position. Figure 9-3 shows when a body is stable and when it is unstable; in addition, it is shown how this stability may be increased by widening the base or lowering the center of gravity.

A third condition of equilibrium is known as neutral equilibrium. Consider objects such as ball bearings, cylinders, and cone-shaped parts. When these objects are moved on a level surface, they neither take on a new position nor return to the original one. When the tipping or rolling of an object neither raises nor lowers the center of gravity, the object is said to be in *neutral equilibrium*.

■ **BALANCING PARALLEL AND ANGULAR FORCES: ADDITION AND SUBTRACTION OF VECTORS**

A force is a push or pull. The effect of a force is either to alter the shape of an object or to prevent other forces from making such a change. Forces may act singly or in combination with other forces. When forces act together, their combined force may be matched by a single force known as the *resultant*. Forces which have a known direction and known magnitude are called *vectors*. Vectors are lines which are drawn to scale to show the amount of force. Arrowheads are used to indicate the direction of force.

Scalar and Vector Quantities

A *scalar quantity* represents a physical quantity that is completely specified by the magnitude of the scalar quantity (expressed by a number) and a unit. An object with a mass of 65 kilograms, an electrical energy rating of 10 amperes, a speed of 42

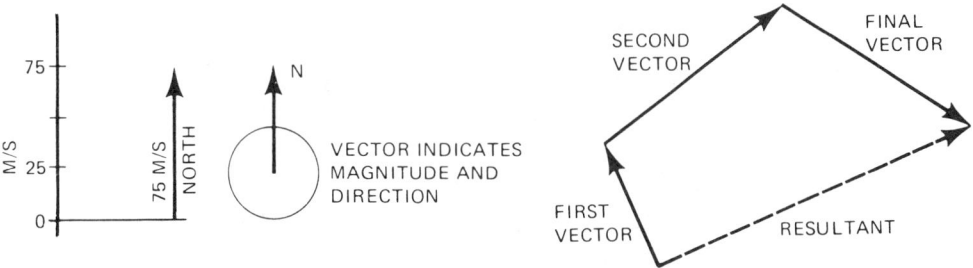

Fig. 9-4 Representing a vector

kilometers per hour, and a volume of 122 cm³ are all examples of scalar quantities. In each instance, the quantities are expressed in *scalar statements*.

For other quantities, the direction of the quantity is as significant as the magnitude. A quantity having both magnitude and direction, as shown in figure 9-4, is called a *vector quantity*. There are three common vector quantities:

1. displacement

2. velocity

3. force

Displacement refers to a change in position. This change is expressed in a *vector statement*. If an automobile is traveling 250 kilometers east from a given point, its displacement at any point can be represented by a line (drawn to scale) to indicate the distance traveled, and an arrowhead at the end of the line to indicate the direction of travel. If other quantities and directions are involved in a problem, it is possible to solve for required values by graphing the given values on a vector diagram or by mathematical computation.

Velocity is represented by the speed of a body and the direction in which it travels. This information makes it possible to determine the path and location of the body at any point in time.

For the vector quantity of *force*, both the direction and magnitude of the force must be stated whenever the effect of the force is to be determined.

Vector problems may be simplified by making a simple drawing of each vector. A vector may be represented by a line that has magnitude and direction. For example, an upward force whose magnitude is 400 kilograms may be represented on a scale of 1 cm equals 100 kilograms. Thus, the vector representing this force is 4 cm long and the vertical direction of the force is indicated by a vertical arrowhead (↑) at the end of the vector.

The tail of each successive vector is placed at the point of the arrowhead of the previous vector. The original length and direction remain unchanged. The resultant is drawn as a vector which connects the tail of the first vector to the head (vertex of the arrowhead) of the last vector quantity (as illustrated in figure 9-4).

Forces Acting in the Same Direction (Parallel Forces)

One important principle concerning forces is that forces acting in the same direction are added to determine the resultant. If two people (A and B) pull the crate shown in figure 9-5 with forces of 60 pounds and 110 pounds, the combined force is 170 pounds (77 kg). The resultant of the two forces acting in the same direction is their sum.

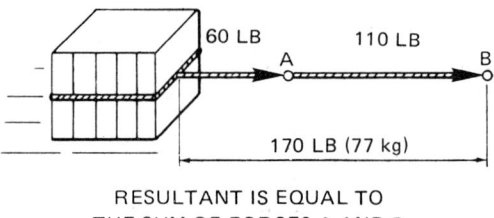

RESULTANT IS EQUAL TO
THE SUM OF FORCES A AND B

Fig. 9-5

The resultant of velocities can be found using the same method as that shown for forces. For example, an airplane traveling north at an air speed of 500 miles per hour has a south tailwind of 50 miles per hour. Since both the plane and the tailwind are traveling in the same direction, the speed of the plane is the sum of its speed (thrust of its engines) and the speed of the wind force. The resultant air speed is 550 miles or 886 kilometers per hour (see I in figure 9-6).

Forces Acting in Opposite Directions (Parallel Forces)

The resultant of forces acting in opposite directions is found by subtracting the forces. The same method applies to velocities. If this same airplane travels due north at the same air speed [500 miles (805 km) per hour] and runs into a 50 mile (80.5 km) an hour headwind from the north, then the airplane's speed is decreased. The resultant speed is the difference between the plane speed and the wind velocity; that is, 500 – 50 or 450 miles (724.5 km) per hour (see II in figure 9-6).

Forces Acting at an Angle to Each Other (Angular Forces)

All forces and velocities do not act in the same direction or in opposite directions. Refer to part III of figure 9-6. If an east wind exerts a wind force of 50 miles per hour,

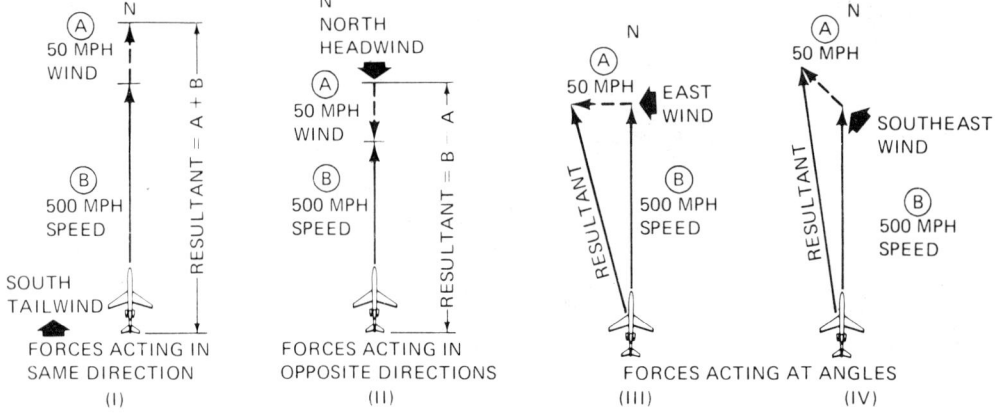

Fig. 9-6

the plane will be blown off its course. If the vectors for the direction and speed of the airplane are drawn, they will form two legs of a right triangle. The resultant is the hypotenuse of a right triangle and may be determined mathematically.

Vector Representation by Graphical Means

It is possible to lay out the triangle on graph paper using the same scale for both vectors (at right angles to each other). The resultant is the third leg of the triangle and can be measured to the same scale as the other two vectors. This is a practical method of finding the resultant in most situations. However, this method is not as accurate as finding the resultant mathematically.

The same resultant speed may be found using the parallelogram method of graphing the result, figure 9-7. In the parallelogram method, the vectors form two sides of a parallelogram. The two opposite and parallel sides of the parallelogram complete the figure. The diagonal of such a parallelogram is the resultant. If the vectors are laid out to the same scale and at the correct angles to each other, the resultant may be measured using the same scale.

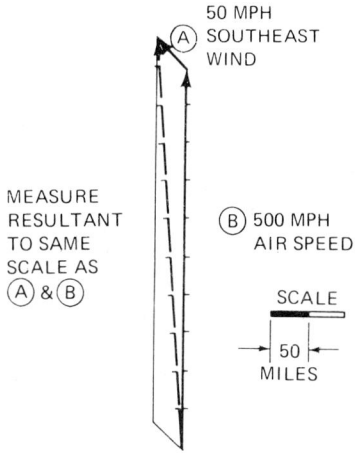

Fig. 9-7

Finally, if the plane continues north at 500 miles (805 km) per hour (part IV of figure 10-6), and there is a southeasterly wind acting on it, an oblique triangle is formed. One vector of the triangle represents the 500 mile (805 km) per hour force of the airplane due north, the second vector at an angle of 45° represents the wind force of 50 miles (80.5 km) per hour in a northwestern direction, and the third leg gives the resultant speed. If a parallelogram of velocities is drawn with both vectors plotted to the same scale, the diagonal may be measured to obtain the resultant speed.

Four types of forces have been considered in this unit: (1) forces acting in the same direction, (2) forces acting in opposite directions, (3) forces acting at right angles to each other, and (4) forces acting at any angle to each other.

The Equilibrant

There are times when a single balancing force is needed to prevent outside forces from moving an object. This balancing force which produces equilibrium is known as an *equilibrant*. The equilibrant is a force that is opposite and equal to the resultant and will balance the vector forces. A drawing which shows the magnitudes and directions of the vectors is known as a *vector diagram*. If a vector diagram consists of two forces (30 and 40 pounds) acting at right angles to each other, figure 9-8, the resultant force required

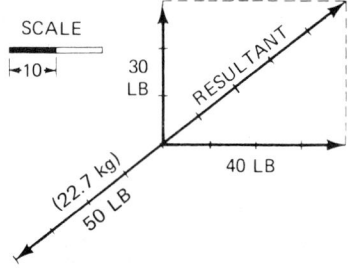

Fig. 9-8 Equilibrant equals a resultant but is opposite in direction

to balance the two forces is 50 pounds (22.7 kg). The equilibrant, therefore, is 50 pounds (22.7 kg).

The term *component* is often used to mean either of two forces which produce a given force as their resultant. Most force problems may be computed or plotted easily when a simple sketch is made to identify the direction and amount of each force. The amount of force is also called its *magnitude*. The point from which the forces act is called the *point of application*. The parallelogram method of determining the resultant is also called the *composition of forces* method.

The four combinations of forces considered in this unit, and the methods of computing or graphing the resultant force, are those that are commonly used to solve problems. The resultant or equilibrant force determined by one method should be checked for accuracy using a second method.

SUMMARY

- An object that is balanced while at rest is statically balanced; when the object is balanced in motion, it is dynamically balanced.

- The center of gravity is the point at which an object is statically balanced.

- A part or mechanism is in stable equilibrium when the center of gravity falls within its base. Unstable equilibrium results when the center of gravity of an object falls outside its base.

- A scalar quantity consists of a numerical value and a unit which completely specify the magnitude of a physical quantity.

- A vector quantity gives both direction and magnitude to completely specify a physical quantity.

- Vectors in a vector diagram originate from a point of application. A vector may be represented by a line whose length is proportional to its magnitude and which extends in a specific direction and terminates in an arrowhead.

- The resultant of two forces acting in the same direction is the sum of the forces.

- The resultant of two forces acting in opposite directions is the difference of the forces.

- The resultant of forces acting at angles to each other can be determined using the triangle of forces method or the parallelogram of forces method.

- The equilibrant is a single balancing force which is equal and opposite to the resultant; the equilibrant produces equilibrium.

ASSIGNMENT UNIT 9 BALANCE AND EQUILIBRIUM — PARALLEL AND ANGULAR FORCES

■ **PRACTICAL PROBLEMS WITH FORCES AND BALANCE**

Dynamic and Static Balance and Stability

For statements 1 to 8, determine which are true (T) and which are false (F).

1. A body that is balanced while at rest is in dynamic balance.
2. The forces that tend to produce rotation are called dynamic balance.
3. Equilibrium refers to a state of balance in which the clockwise and counter-clockwise rotation are equal.
4. Uneven wear and excessive vibration of moving parts may be due to dynamic unbalance.
5. A bicycle wheel, automobile wheel, or pulley may be dynamically balanced by locating the points where weight may be added or removed, as needed.
6. A solid plate is stable when the center of gravity falls outside its base.
7. Car designers try to concentrate the weight and center of gravity of cars as low as possible for safety and stability.
8. A spherical ball bearing and objects that are cylindrical or cone-shaped are in neutral equilibrium because the center of gravity remains constant when they are rolled.

Balancing Parallel and Angular Forces (Scalar and Vector Quantities)

1. Define the following and give an example to illustrate each vector term.
   a. scalar quantity
   b. vector quantity
   c. displacement

Select the correct word or phrase to complete statements 2-6.

2. Forces which are combined and act together may be matched by (a single force) (two or more forces).
3. The resultant of two forces pulling in the same direction is the (difference between) (product of) (sum of) the two forces.
4. The resultant of two forces pulling in opposite directions is the (product of) (sum of) (difference between) the two forces.
5. In a vector diagram of two forces acting at a right angle to each other, the hypotenuse of the right triangle formed is the (equilibrant) (resultant).

6. The equilibrant is a force that (is equal and opposite) (is unequal and opposite) (has no relationship) to the resultant.

7. The airline distance between two points is 2040 miles. An airplane traveling at an air speed of 500 miles an hour due south runs into headwinds of 75 miles per hour due north. How many hours does it take the airplane to travel this distance?

8. Determine the resultant force in problems A, B, and C of figure 9-9. Check each problem using a second method. Show how each answer is determined.

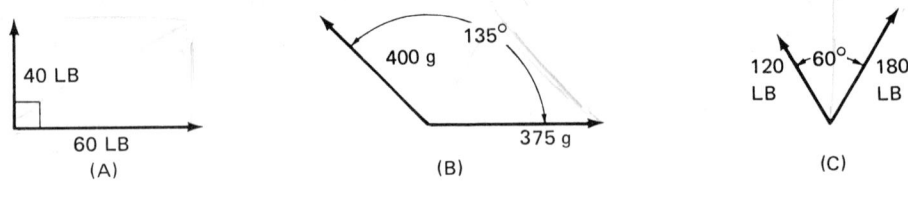

Fig. 9-9

9. A guy wire extends at an angle of 30° from the side of a building. The pull on this wire, which helps to support a sign, is 400 pounds. If the angle of the guy wire is changed to 45°, will it require a greater or a lesser pull? How much?

10. Compute the force required to pull a 400-kilogram object up a 30° incline at a constant speed. Assume the forces are applied parallel to the plane. Neglect friction.

11. Convert the force required for problem 11 to its customary unit equivalent.

Parallel balancing ways for checking wheel trueness and distribution of weight (Courtesy of DoAll Company)

Positioning balancing weights to overcome centrifugal forces at high speeds (Courtesy of DoAll Company)

Techniques of correcting out-of-balance centrifugal forces in grinding wheels for safety, to eliminate vibration, and to produce high dimensional accuracy and fine surface textures

Unit 10 Gravitation, Motion, and Mechanical Movements

Aristotle conducted many experiments with falling bodies and kept careful records of his findings. Although his work was later found to be in error, the importance of Aristotle's work lies in the fact that Galileo based his experiments on Aristotle's experiments. In the early 17th century, Galileo proved that the weight of a body has little to do with the speed at which a body falls. Two different bodies, such as a feather and a coin, fall at the same rate when dropped in a vacuum.

■ FUNDAMENTAL LAWS OF GRAVITATION

Galileo's Work with Falling Bodies

Distance Covered by Objects Rolling Down Inclined Surfaces. While experimenting with inclined surfaces, Galileo found that the distance covered by an object rolling down an incline is equal to the distance it travels in the first second multiplied by the square of the time traveled. If a one-inch diameter ball rolls down an inclined surface one foot (0.30 m) in one second, it will travel four feet (1.22 m) in two seconds, nine feet (2.75 m) in three seconds, 16 feet (4.88 m) in four seconds, and so on, figure 10-1.

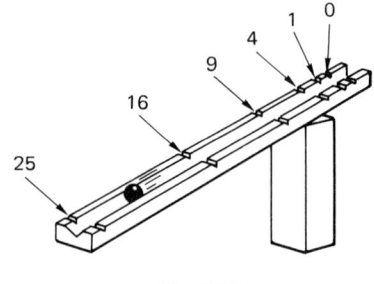

Fig. 10-1

The distance traveled can be expressed by the following formula:

$$D = d \times T^2$$

where D = total distance that the object travels, in feet
          d = distance covered for the first interval of time, in feet
          T = time, in seconds

By substituting values from the above example, $D = 1 \times 1^2$ or one foot (0.30 m) the first second; 1×2^2 or four feet (1.22 m) in two seconds; 1×3^2 or nine feet (2.75 m) in three seconds; and 1×4^2 or 16 feet (4.88 m) in four seconds.

Distance Covered by Freely Falling Objects. If an object falls straight down instead of rolling down an incline, the distance the object falls is again equal to the distance traveled in the first second multiplied by the square of the time. It is known, however, that freely falling bodies (neglecting any losses caused by the friction of the air) always fall 16 feet (4.88 m) the first second.

Thus, the distance traveled by a freely falling body is given by the formula:

$$D = 16T^2 \qquad \text{or,} \qquad D = 4.88T^2$$

where D = total distance, in feet   where D = total distance, in meters
          T = time, in seconds   T = time, in seconds

Gravitation, Motion, and Mechanical Movements   83

In one second, then, an object falls 16 × 1² or 16 feet; in two seconds, 16 × 2² or 64 feet; in three seconds, 16 × 3² or 144 feet; in 10 seconds, 16 × 10² or 1600 feet; and in 30 seconds, 16 × 30² or 14 400 feet. Similarly, in SI metrics, an object falls 4.88 × 1² or 4.88 meters in one second; 4.88 × 2² or 19.52 m in two seconds; 4.88 × 3² or 43.92 m in three seconds, etc.

Newton's Law of Gravitation

Many years after Galileo, Sir Isaac Newton continued experimenting with gravity and its effects on falling bodies. In simplified form, Newton's *law of gravitation* means that all objects in the universe attract all other objects and the closer two bodies are to each other, the greater is the attraction.

Newton also determined that the masses of the bodies influence the force of this attraction. All objects are attracted to the earth because its mass is larger than any object on it. The attractive force that pulls all things toward the center of the earth is called *gravity*. In addition, all falling objects gain speed at a uniform rate which is not affected by the weight of the object.

Velocity of Freely Falling Objects. A freely falling object starting at an initial velocity of zero reaches a speed of 32 ft/s (9.76 m/s) at the end of the first second. For each succeeding second, the free-falling object increases its speed by 32 ft/s. Therefore, the velocity of a free-falling object at any time (disregarding air resistance) is equal to 32 ft/s (9.76 m/s) multiplied by the time in seconds.

Expressed as a formula, the velocity is:

$$V = 32T \quad \text{or,} \quad V = 9.76T$$

where V = velocity, in feet/second
T = time, in seconds.

where V = velocity, in meters/second
T = time, in seconds.

Starting from rest, a freely falling object has a velocity at the end of one second of 32 × 1 or 32 ft/s; at the end of two seconds, 32 × 2 or 64 ft/s; at the end of three seconds, 32 × 3 or 96 ft/s; at the end of 10 seconds, 32 × 10 or 320 ft/s; and so on. In SI metric, freely falling object has a velocity at the end of one second of 9.76 × 1 or 9.76 m/s at the end of two seconds, 9.76 × 2 or 19.52 m/s, etc.

■ COMMON CONCEPTS ABOUT MOTION

A body may experience three kinds of motion. The body may move in straight or curved lines, it may rotate about its axis without moving from a fixed position, or it may vibrate. The term *motion* refers to a change of position. This motion may be either absolute or relative, figure 10-2. In *absolute motion*, the body moves away from a fixed point of reference (as shown at A in figure 10-2). On the other hand, if the body moves away from another point or object that is also moving, the motion, as shown in figure 10-2B, is *relative motion*. A body is considered to be at *rest* when its position remains unchanged with respect to some given point or object.

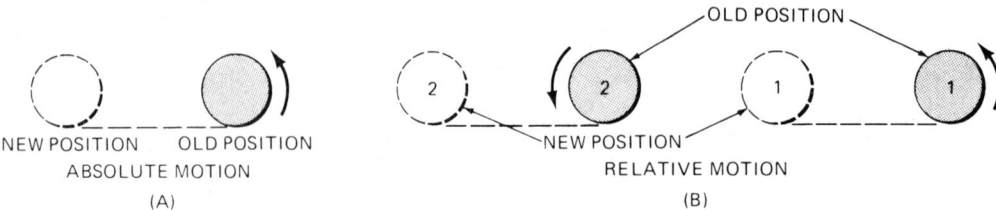

Fig. 10-2

When bodies are in motion, two terms are used to describe the motion: speed and velocity. *Speed* is defined as the distance traveled divided by the time of travel. *Velocity* includes both speed and direction. Both terms are used interchangeably in this and other units.

Speed and Acceleration

The total distance covered by a moving object, divided by the time required to cover the distance, defines a quantity called the *average speed*. The term *constant speed* is applied to motion at unchanging speed; for example, a machine may produce wire or paper or plastic at a constant speed, such as 1000 feet (305 m) per minute; this speed is also the average speed. If a car travels 60 miles (97 km) in two hours, its average speed is 30 miles (48 km) per hour. This speed probably was not a constant speed, but varied between 0 and 70 miles (113 km) per hour. The rate at which speed changes is called *acceleration*. *Positive acceleration* implies a gain of speed; a loss of speed is *negative acceleration*. Numerically, acceleration equals the change in velocity divided by the time in which the change occurs. If a car increases its speed from 15 miles (24 km) per hour to 25 miles (40 km) per hour in two seconds, its acceleration is 5 miles (8 km) per hour per second. If an object has *constant acceleration*, it is changing speed steadily, either speeding up or slowing down. A body that maintains a steady unchanging speed has *zero acceleration*.

Average and Final Speeds and Accelerated Motion

Average Speeds. When a body accelerates so that it either gains or loses speed uniformly, the *average speed* is the sum of the initial speed and the final speed divided by two. If an automobile travels at a speed of 30 miles (48.3 km) per hour, figure 10-3, and gains speed at a constant rate until it reaches 60 miles (96.6 km) per hour, then:

Average speed = (initial speed + final speed)/2

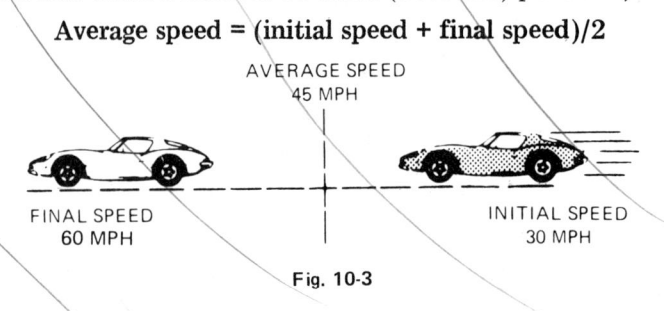

Fig. 10-3

Substituting values in this expression yields:

$$\frac{30 + 60}{2} = 45 \text{ miles per hour} \qquad \text{or,} \qquad \frac{48.3 + 96.6}{2} = 72.5 \text{ km per hour}$$

The same automobile climbing a steep hill may lose speed uniformly from 60 to 30 miles (96.5 to 48.3 km) per hour. Again, the average speed is 45 mph (72.5 km/h) as determined from the same formula.

Final Speeds. The *final speed* of a uniformly accelerating body is the speed at a given instant. Starting at rest, the final speed equals the elapsed time multiplied by the acceleration. If an object starting at rest gains speed down an incline at the rate of 10 feet per second each second, its final speed after half a minute is 30 seconds × 10 ft/s/s = 300 ft/s. The final speed of a body that loses speed uniformly is found in the same manner. The final speed (S_f) is equal to the product of the uniform acceleration (A) × time (T) plus the initial speed of the object (S_i), figure 10-4.

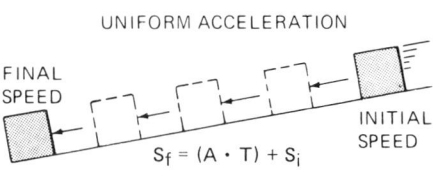

Fig. 10-4

$$S_f = (A)(T) + S_i$$

Substitute values from the preceding example in the expression to obtain the final speed:

$$S_f = \frac{10 \text{ ft/s}}{s} \times 30 \text{ s} + 0 = 300 \text{ ft/s} \qquad \text{or,} \qquad S_f = \frac{3.05 \text{ m/s}}{s} \times 30 \text{ s} + 0 = 91.5 \text{ m/s}$$

Care must be taken to insure that the units of measure used in acceleration problems are practical and consistent. An aid to understanding and solving acceleration problems is the drawing of a simple sketch of the given conditions. Once the sketch is drawn, the values that must be found can be visualized and all values can be labeled.

Uniform Motion. *Uniform motion* is steady motion in a straight line at a constant speed. Uniform motion (distance) is equal to the product of the velocity (V) and time (T). A train moving at a uniform speed of 10 feet per second for 12 seconds travels a distance of:

$$D = (V)(T) = (10 \text{ ft/s})(12 \text{ s}) = 120 \text{ ft or } 36.6 \text{ meters}$$

Uniform acceleration or *deceleration* indicates a constant increase or a constant decrease in speed, respectively. The velocity gained due to a constant acceleration is equal to the product of the acceleration and time. Conversely, the velocity lost by constant deceleration is the product of the rate of deceleration and time.

$$V_g = (A)(T)$$
$$V_l = (D)(T)$$

where V_g = Velocity gained
V_l = Velocity lost
A = Acceleration
D = Deceleration
T = Time

Thus far, definitions have been given for the terms used to describe the kinds of motion at different speeds. These conditions are now studied in terms of the basic laws which affect motion. It is important to know these laws because they are applied daily in industry, on the farm, in business, and in the home. The three laws to be studied are known as *Newton's Laws of Motion* because of the work of Sir Isaac Newton and his explanation of these laws around 1687.

■ NEWTON'S FIRST LAW OF MOTION

The Behavior of Bodies at Rest and in Motion

Very simply interpreted, the first law states two truths.

1. Every body that is at rest tends to remain at rest unless an external force produces a change in the state of rest.
2. Every body in motion remains in motion unless an external force causes the body to change that motion.

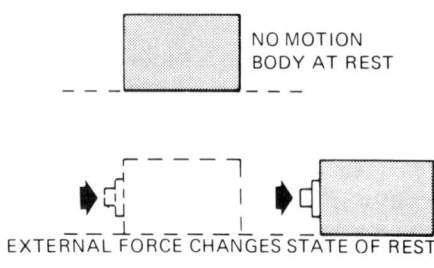

Fig. 10-5

Whenever bodies at rest or in motion are discussed, the term inertia is often used. *Inertia* describes the tendency of a body to remain at rest. Inertia also relates to a body which is in motion and continues its motion at the same speed and direction unless an outside force is applied.

The term *rotational inertia* is used with objects that spin around an axis and have a tendency to continue to rotate around the same axis. The principle of rotational inertia is used in the design of stabilizing devices such as the gyrocompass and the gyrostabilizer in airplanes.

■ NEWTON'S SECOND LAW OF MOTION

Forces Affecting Motion

A body in motion has a certain quantity of motion which is called its *momentum*. Two factors control the momentum of a body at any given time: the body's *mass* and its *velocity*. Thus, momentum may be expressed as the product of the mass and the velocity of the body.

Newton's second law states that any change in the quantity of motion (momentum) is directly proportional to the force acting upon a body. This means that the less the time required to decelerate an object, the greater is the force acting on the object. Newton's second law is used in common applications such as the heavy flywheel of a gas or diesel engine.

A small force is used on engines to gradually build up the speed of the flywheel. Once the desired flywheel momentum is reached, little effort is required to keep the flywheel (or other heavy part) in motion. The energy in the momentum of a heavy flywheel can be recovered with a great increase in force by cutting the speed abruptly.

Gravitation, Motion, and Mechanical Movements

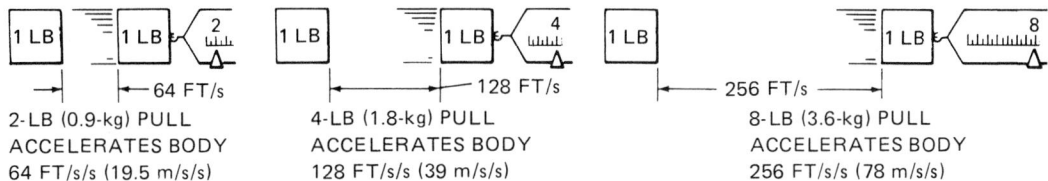

2-LB (0.9-kg) PULL
ACCELERATES BODY
64 FT/s/s (19.5 m/s/s)

4-LB (1.8-kg) PULL
ACCELERATES BODY
128 FT/s/s (39 m/s/s)

8-LB (3.6-kg) PULL
ACCELERATES BODY
256 FT/s/s (78 m/s/s)

Fig. 10-6 Acceleration is proportional to force

This type of situation is found in a power press for stamping or forming materials. As the clutch mechanism transfers the rotary motion (momentum) of the flywheel to a reciprocating motion of the press ram, a tremendous force is exerted on the forming, cutting, or bending dies. These dies, in turn, cut and shape the material.

If a force is applied in the same direction as the motion of a body, the motion of the body increases. The direction in which the applied force acts and the magnitude of the applied force determine the amount of change in the momentum of the body.

Stated in another way, Newton's second law of motion says that the force required to accelerate an object is proportional to the mass of the object and to the acceleration produced, figure 10-6.

$$\text{Required Force} = \text{Mass} \times \text{Acceleration}$$

$$F = M \times A$$

■ NEWTON'S THIRD LAW OF MOTION

Action and Reaction

The third law of motion states that every action has an equal and opposite reaction. That is, if object A pushes or pulls on object B, then, at the same time, B pushes or pulls on A with the same amount of force.

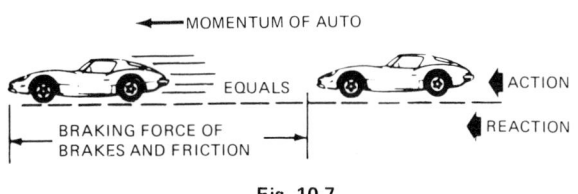

Fig. 10-7

As an example of this law in action, consider that work must be fastened to machines with sufficient force to overcome the force exerted by a cutting tool on the work. A slab of steel is held in a milling machine vise to resist the forces of the cutter teeth when removing metal. The forward driving force of a jet plane is produced by the backward thrust of the jet engine.

■ CENTRIFUGAL AND CENTRIPETAL FORCES

The inertia of a body tends to cause the body to travel in straight lines. The sparks (particles of material) thrown off grinding wheels fly from the wheel on paths which are tangent to the circumference of the wheel. A force known as *centrifugal force* tends to cause rotating objects to fly apart.

Considerable attention is given to the action of centrifugal force in grinding wheels. That is, the bonding agent that holds the abrasive particles on the wheel must be stronger than the forces which tend to make the revolving wheel fly apart at high speeds. For this reason, the speed of a grinding wheel should not exceed the safe surface speed limit specified by the manufacturer. Centrifugal force increases with speed.

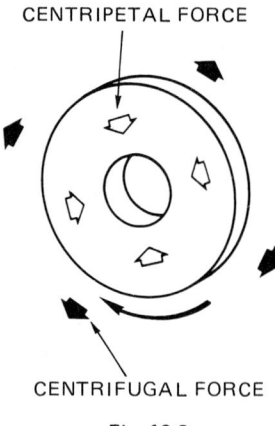

Fig. 10-8

The principles of centrifugal force are used in the design of centrifuge-type machines. Some centrifuges are used to separate chemicals; others are used to remove impurities in metals by centrifugal casting processes. Centrifugal force principles are also used in common appliances such as clothes dryers and in devices to control motor speeds and accelerate moving machines.

Centripetal force causes an object to travel in a circular path. This action is caused by the continuous application of forces which tend to pull the object to the center. In other words, the inward force which resists the centrifugal force is called the centripetal force. The centripetal force of objects spinning at a constant rate produces an acceleration toward the center which is equal and opposite to the centrifugal force.

The materials used in the construction of rapidly moving machine parts and mechanisms must be structurally sound and strong enough to provide the centripetal force required to hold the parts to a circular path. At the same time, the materials must be able to withstand the centrifugal force which tends to pull the parts apart.

Centripetal force and centrifugal force can be measured using the same formula:

$$C = \frac{W \times V^2}{g \times r}$$

where C = centrifugal or centripetal force
W = weight of the object
V = the speed or velocity of the object
g = acceleration caused by gravity
r = the radius of the path of the object

■ MECHANICAL MOVEMENTS

Motion and the basic laws which affect motion are important considerations because of the numerous applications of these principles to produce work through mechanical devices. There are two primary mechanical motions: rotary and rectilinear. These terms suggest that *rotary motion* is a circular movement around a center line and *rectilinear motion* is a straight line motion. For either rotary or rectilinear motion, it is possible, with added mechanical devices, to produce other forms of motion such as intermittent motion and reciprocating motion.

Transmitting Motion

Rotary Motion. The motion that is commonly transmitted is rotary motion. This type of motion may be produced with hand tools or power tools. Rotary motion is required to drill holes, turn parts in a lathe, grind tools, mill surfaces, cut through materials, or drive a generator or fan belt.

Rectilinear Motion. The feed of a tool on a lathe, the cutting of steel on a power saw, or the shaping of materials are all situations in which rectilinear or straight line motion produces work. In each of these situations a part or mechanism is used to change rotary motion to straight line motion. The screw of a micrometer, the threads in a nut, and the cross-feed screw on a lathe are still other applications where the direction of motion is changed from rotary to rectilinear.

Harmonic and Intermittent Motion. Any simple vibration, such as the regular back-and-forth movement of the end of a pendulum, is simple *harmonic motion*. However, many manufacturing processes require *intermittent* or irregular motion. For example, the fast return stroke of a power hacksaw or shaper ram is desirable because no cutting is done on the return stroke. Therefore, as more time is saved in returning the blade or cutting tool to the working position, the less expensive is the operation.

A device known as a *cam*, figure 10-9, is commonly used to change a uniform motion to an intermittent or irregular motion. The cam may rotate or oscillate. The cam has a surface or groove formed in such a way that it gives a desired motion to another part.

The combinations of rotary and rectilinear motion obtainable are unlimited because of the large variety of parts such as gears, cams, pulleys, screws, links, and belts which can be combined in many arrangements.

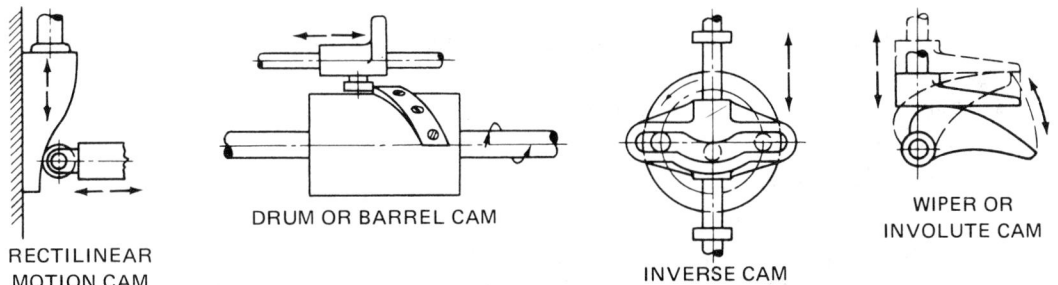

Fig. 10-9

Computing Mechanisms

Mechanical movements and mechanisms are also incorporated in mathematical computation machines. Mechanical movements are used to make calculus computations, determine algebraic functions, and establish physical quantities involving angles and trigonometric functions. In each of these cases, the mechanism may be classed as an *analog computer.* The term *analog* means a relationship or similarity between two different things.

A specific length, distance, angle, or movement of a mechanism may represent time, distance, or a volume such as a liter or gallon per hour. Another mechanism may represent a specific force, work, or power output. The movement of one device is analogous to the information provided by another component of the mechanism.

Mechanical movements also are combined with electrical and electronic equipment to yield devices suitable for other types of computational applications. For example, a speedometer cable on an automobile is rotated by a gear (driving force) on the cable which meshes with a gear in the transmission. The cable, in turn, causes the rotation of a permanent magnet inside an electric coil. The amount of electric current generated (output) is proportional to the rate of change in the movement of the vehicle. Speed is measured by reading the position of the pointer with relation to the values calibrated (in miles per hour or kilometers per hour) on the face of the meter (speedometer).

SUMMARY

- All bodies in the universe attract all other bodies.
- The mass of a body influences gravitation.
- All freely falling objects gain speed at a uniform rate of 32 feet per second per second regardless of the weight of the objects.
- Motion is the change of position of a body.
- Motion may be absolute or relative.
- A body has positive acceleration when it gains speed uniformly; a body with negative acceleration loses speed uniformly.
- The average speed of a uniformly accelerating body is equal to (initial speed + final speed) ÷ 2.
- Newton's Three Laws of Motion deal with inertia, momentum, and action and reaction.
  - The first Law of Motion states that a body at rest will remain at rest unless disturbed by an outside force.
  - The Second Law states that the change in the momentum (F) of a body is proportional to the mass (M) and acceleration (A) of the body.
    $$F = M \times A$$
  - The Third Law states that for every force or action there is an equal and opposing force or action.
- Centrifugal force tends to cause a rotating object to fly apart; centripetal force tends to resist centrifugal force.
- Centrifugal or centripetal force is determined by the formula,
$$C = \frac{W \times V^2}{g \times r}$$

- Rotary motion is used in hand tools, machines, and mechanical devices to do work by moving parts or tools with a circular motion.
- Rectilinear motion is straight line motion which may be intermittent, uniform, or any combination of these.

ASSIGNMENT 10 GRAVITATION, MOTION, AND MECHANICAL MOVEMENTS

■ PRACTICAL PROBLEMS IN GRAVITATION, MOTION, AND MECHANICAL MOVEMENTS

Early Experiments with Gravitation

Select the correct word or phrase to complete statements 1 to 6.

1. All objects in the universe (are) (are not) attracted to each other.
2. The (mass) (shape) of a body influences gravitation.
3. All freely falling objects gain speed at (an irregular) (a uniform) rate.
4. For an object rolling down an incline, the total distance (D) traveled is equal to (d × t^2) (d × t) (d × t^3), where d = distance and t = time.
5. As two bodies draw closer to each other, the attraction between them (decreases) (increases).
6. (Gravity) (Motion) is the attractive force that pulls all things toward the center of the earth.
7. Determine the total distance traveled by round parts as they move down an inclined, grooved conveyor at intervals of one second for each of five seconds. Use speeds of 3 inches, 4 inches, 5 inches, 6 inches, and 8 inches per second, respectively, and the formula D = (d)(t^2). Record the distances in a table similar to the one shown.

Time Intervals (s)	Speed (in/s)				
	3	4	5	6	8
1					
2					
3					
4					
5					

8. Change the speed of the parts in problem 7 to SI metric (cm) equivalents. Compute and record the speeds for time intervals of one and two seconds.

Time Intervals (s)	Speed (cm/s)				
	3	4	5	6	8
1					
2					

Concepts of Motion

Complete statements 1 through 8.

1. _____ refers to a body that changes position.
2. Three common types of motion are: _____, _____, and _____.
3. The total distance covered divided by time is the _____.
4. A gain or loss in speed is referred to as _____.
5. The average speed is equal to _____.
6. The uniform acceleration of an object × time + the initial speed is the _____.
7. In uniform motion, the distance covered is equal to _____ multiplied by _____.
8. The velocity an object loses by constant deceleration is equal to _____ multiplied by _____.
9. Determine the average speeds of a mechanism having the initial and final speeds shown in the table.

	Initial Speed (rpm)	Final Speed (rpm)	Average Speed (rpm)
A	16	40	
B	225	675	
C	40	16	
D	675	225	

10. A series of motors accelerates uniformly in one second at the rate indicated in the table. Determine the final speed of each motor.

Motor	Initial Speed (Revolutions)	Uniform Acceleration (rps)	Time to Reach Final Speed (s)	Final Speed (Revolutions)
A	At rest	20	10	
B	At rest	35	8	
C	20	65	30	
D	At rest	420	4 1/2	

Newton's Three Laws of Motion

Select the correct word or phrase to complete statements 1 to 5.

1. (Momentum) (Inertia) (Acceleration) is the tendency of a body to remain at rest, or, if in motion, to continue its motion at the same speed.

Gravitation, Motion, and Mechanical Movements

2. A body at rest (tends to remain at rest) (moves) when pulled upon by two equal and opposite forces.
3. The faster a body decelerates, the (smaller) (larger) the forces acting upon it.
4. The change in momentum of a body is determined by (the magnitude and direction of an applied force) (the inertia of the body) (the acceleration).
5. For every force, there is an (equal) (unequal) and (complementary) (opposing) force.
6. Give three examples from industry that show how every body remains at rest or continues in motion unless an external force produces a change.
7. List two industrial applications of Newton's second law of motion in which the force required to accelerate an object is proportional to the mass of the object and the acceleration produced.
8. List two industrial or commercial applications of Newton's third law of motion dealing with actions and reactions.

Centrifugal and Centripetal Force

For statements 1 to 5, determine which are true (T) and which are false (F).
1. Centrifugal force decreases with speed and weight.
2. The centrifugal force of a revolving body causes the body to resist flying apart.
3. Centripetal force is an inward force tending to resist the centrifugal force.
4. At constant speed, the centripetal and centrifugal forces are equal.
5. The centripetal force decreases with speed and weight.
6. Explain briefly why a chipped or cracked abrasive wheel should not be used.
7. What safety precautions does the abrasive wheel manufacturer build into an abrasive wheel? Explain the answer in terms of the centripetal and centrifugal forces.

Mechanical Movements

1. Name two mechanical devices which use rotary motion.
2. Describe what is meant by rectilinear motion.
3. List three industrial applications of rectilinear motion.
4. What is the difference between harmonic motion and intermittent motion?
5. Give an example of each type of motion: harmonic and intermittent.
6. A mechanical/electrical device is to be designed to compute the rate of change of one quantity with relation to another quantity.
    a. Identify a common application of the combination of a mechanical movement and an electrical device to compute (measure) changes of speed against time.
    b. Sketch the mechanism and label the major parts.
    c. State briefly how the mechanism works.

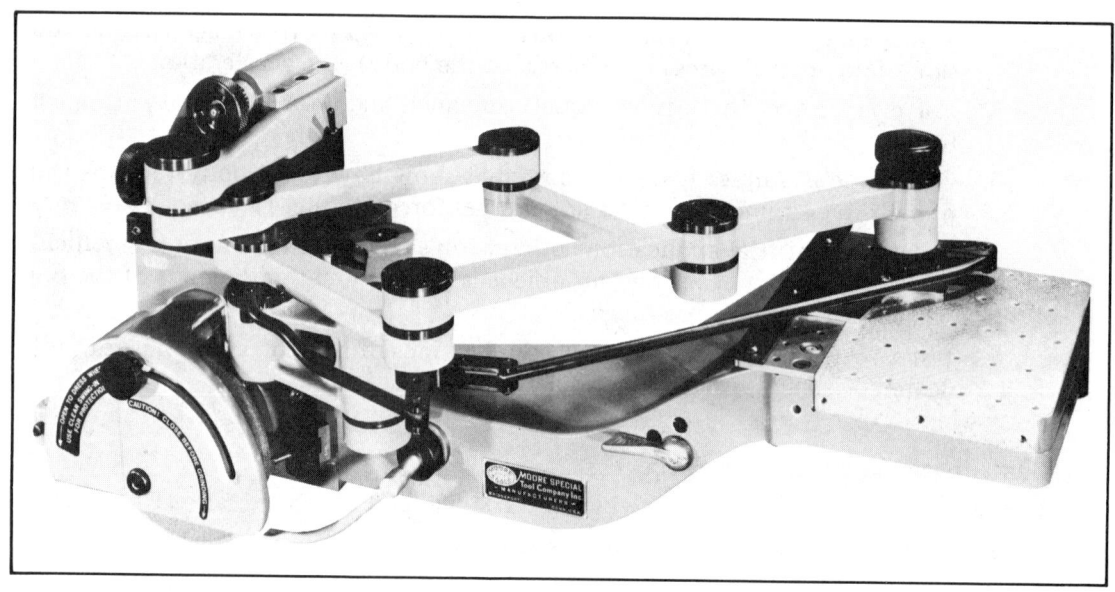

Lever systems for a pantograph attachment to follow a profile template and precisely reproduce the form to a required smaller size (Courtesy Moore Special Tool Company, Inc.)

Unit 11 Simple Machines: Levers

A machine is a mechanical device which uses given forces and directions to produce required work. The function of a machine is to change the magnitude or intensity of a force, the direction of a force, or the speed resulting from a force. Machines may be simple or they may be complex and involve many parts and mechanisms moving in one or more directions and at varying speeds. Examples of simple machines include a screwdriver, a wrench, and a hammer. Simple machines may be used alone or they may be combined, depending upon the kind of work and motion desired. The lever and the inclined plane are the two basic machines. The wheel and axle, the pulley, and the wedge and screws are adaptations of these basic machines.

This first unit on simple machines deals with the lever, its importance, and its application in industry. Attention is focused on the diagraming of levers, the principles of levers, the mechanical advantage of levers, and how a small force may be used to control or develop a larger force.

■ CHARACTERISTICS OF LEVERS

The lever is one of the oldest and one of the simplest of all mechanical devices. A simple *lever* is nothing more than a rigid bar which turns around a fixed point. A straight steel bar used to tip up one end of a heavy crate is an example of a lever used as a simple machine. One end of the lever is placed under the crate. A block of wood located under the steel bar near the point of application is the fixed point around which the lever revolves as force is applied to the opposite end.

The fixed point (in this case, the block under the steel bar) is called the *fulcrum*, the crate is the *resistance* and the force is the *effort*. The fulcrum, resistance, and effort may be given or they must be computed for all lever problems. In addition, two other values are needed, the *effort distance* and the *resistance distance*, figure 11-1. Each value represents a distance between the fulcrum and the point of application of the effort and resistance, respectively.

Levers are usually diagramed to simplify the solution of problems and to help in visualizing the known and unknown values. The parts of a lever diagram are illustrated. The letters commonly used to indicate given values are shown for each of the three types of levers covered. Regardless of the type of lever, the terms used have the same meaning.

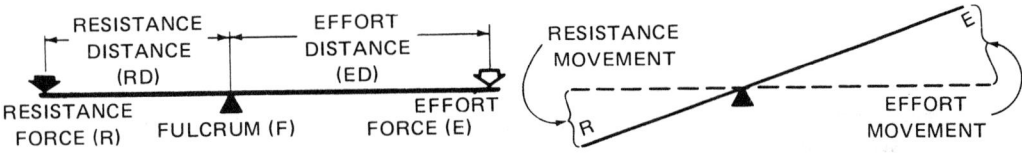

Fig. 11-1 Parts of a lever

■ TYPES OF LEVERS

Levers of the First Class

Levers are classified according to the location of the effort and resistance forces with respect to the pivot point. For a *first class lever* the fulcrum or pivot point is located between the effort and resistance forces, figure 11-2. A first class lever increases the effort force and reverses the direction of motion. In other words, when a downward force is exerted on this type of lever, the force is applied upward against the resistance. A crowbar, a claw hammer and a pickaxe are all common examples of this class of lever. All of these tools use a smaller force to control or develop a larger force.

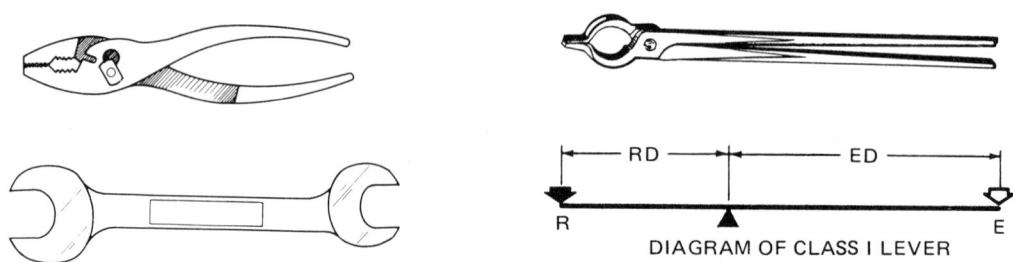

Fig. 11-2 Examples of Class 1 levers

Levers of the Second Class

Levers of the *second class* are also used to gain force, but the fulcrum is located on one side of both the effort and resistance forces, and the direction of motion is the same. A wheelbarrow is one common example of a Class 2 lever, figure 11-3. Another example is a beam used to lift a heavy structure. In each case, the fulcrum is at one end of the lever and the load (R) is located between the fulcrum and the effort force (E).

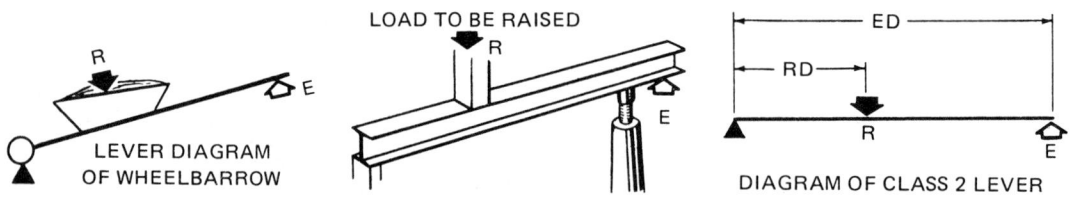

Fig. 11-3 Examples of Class 2 levers

Levers of the Third Class

Levers of the *third class* increase speed or distance, but not force. The fulcrum of a Class 3 lever is at one side of the two forces just as it is for a Class 2 lever. The effort force, however, is applied between the pivot and the resistance, figure 11-4. Thus, the

Simple Machines: Levers 97

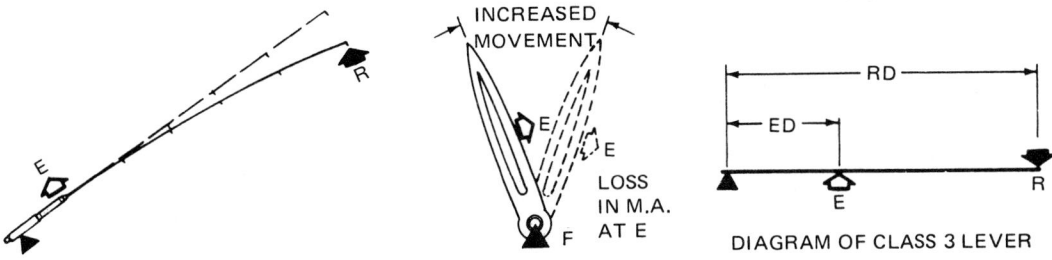

Fig. 11-4 Examples of Class 3 levers

distance between the pivot and applied force (effort distance) is less than the distance between the pivot and the load (resistance distance). The direction of movement is, of course, the same for both forces.

The fishing pole shown in figure 11-4 is an example of a Class 3 lever. The fulcrum is the point at which the pole is held against the body. The force applied by the fisherman's hands (E) increases the distance of movement at the tip of the rod (R).

Many indicating mechanisms make use of third class levers. For example, the pointer of a precision gauge shows a large movement on the dial face when a small force is applied in the working mechanism.

■ MOMENTS OF FORCE

Both the effort force (E) and the resistance force (R) tend to turn the lever around the fulcrum. This tendency to revolve is called the *torque* or *moment of force*. The moment of force is equal to the force multiplied by the distance from the fulcrum. The (R) moments = R × RD and the (E) moments = E × ED. In figure 11-5, a resistance (R) of 10 pounds acts 10 feet from the fulcrum and tends to rotate the lever counterclockwise. The moment of force for (R) = (10 feet) × (10 pounds) = 100 pound-feet (3 m × 4.5 kg = 13.5 kg·m). Opposing the (R) moment of force is the effort (E) of 25 pounds acting four feet from the fulcrum. The (E) moment of force = (25 pounds) × (4 feet) = 100 pound-feet (11.4 kg × 1.2 m = 13.7 kg·m). When both the (E) and (R) moments are equal, the lever is balanced. This condition of balance indicates that the lever does not move. For a balanced lever, the *Law of Moments* states that the sum of the moments which tend to rotate a lever clockwise is equal to the sum of the moments which tend to turn the lever counterclockwise.

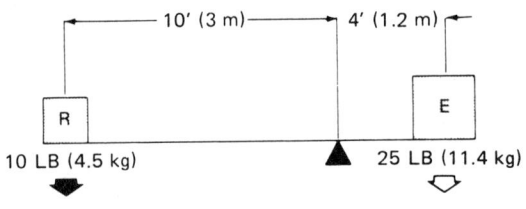

Fig. 11-5

In the previous example, the weight of the lever was neglected. When the weight of the lever is to be considered, the weight is assumed to act at its center of gravity.

MECHANICAL ADVANTAGE OF LEVERS

Mechanical Advantage of Force

Levers are used principally to control larger and heavier forces. The control depends upon the distance of the forces from the fulcrum. As shown in figure 11-6, the 100-pound weight placed on a 10-foot beam may be raised by placing a 25-pound weight eight feet from the fulcrum.

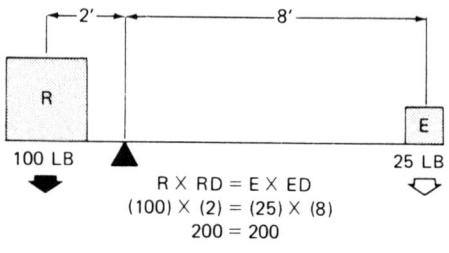

Fig. 11-6

$$R \times RD = E \times ED$$
$$(100)(2) = (25)(8)$$

The (R) moments of force (200 pound-feet) = the (E) moments of force. Thus, a lighter force placed at a greater distance from the fulcrum is used to control a heavier force. As a force is moved farther and farther from the fulcrum, the moments increase. A 100-pound force placed one foot (45.4 kg force at 0.3 m) from the fulcrum produces a moment of 100 pound-feet (13.62 kg·m). At a distance of two feet from the fulcrum, the moment equals 200 pound-feet.

The moment depends on the force and the distance through which this force acts. Therefore, it is possible to use any combination of force and distance having the same product as the moment of force to be overcome. For the previous example, instead of a 25-pound (11.4-kg) effort placed at a distance of eight feet (2.4 m) from the fulcrum, a 50-pound (22.7-kg) effort at four feet (1.2 m), or a 40-pound (18.2-kg) effort at five feet (1.5 m) may be used. Furthermore, a combination of two or more forces, equal to the same moment of force, may be used.

The mechanical advantage of force (MA_f) equals the resistance force (R) divided by the effort force (E):

$$MA_f = \frac{R}{E}$$

If a 50-pound effort overcomes a 100-pound resistance, the mechanical advantage of force is 100 ÷ 50 = 2. Similarly, for equal kilogram values of 45.4 kg ÷ 22.7 kg, the mechanical advantage of force is 2.

Mechanical Advantage of Speed or Movement

The distance that either the effort or the resistance moves and its speed depend upon the position of the fulcrum. When a small force is used to move a larger force, the distance through which the (E) moment moves is greater than the distance of the (R) moment. Figure 11-7 shows that the effort arm (ED) is twice as long as the resistance arm (RD);

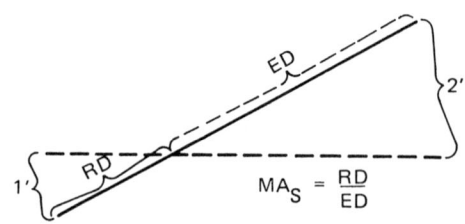

Fig. 11-7

therefore, the resistance moment moves one foot as compared with two feet for the (E) moment. In addition, the speed of the effort arm is twice that of the resistance arm because the effort arm moves twice as far in the same time. The mechanical advantage of speed (MA_S), however, is only one-half. As a result, the gain in force is balanced by the loss in having to move a greater distance to do the same work. The mechanical advantage of speed is expressed by the following formula.

$$MA_S = \frac{RD}{ED}$$

Levers may be compared by examining the ratio of resistance to effort (to find the mechanical advantage of force), or the ratio of resistance distance to effort distance (to find the mechanical advantage of speed).

■ INDUSTRIAL APPLICATIONS OF LEVERS

One of the best known examples of a lever is the wrench. Tremendous force may be applied when tightening or loosening nuts with wrenches. This force is due to the fact that the effort force can be applied at a greater distance from the fulcrum.

Hammers of different types are also examples of the lever. Straps used for clamping work solidly to machine tables depend on leverage for increased forces. Vises also use the principles of the lever to apply force, figure 11-8.

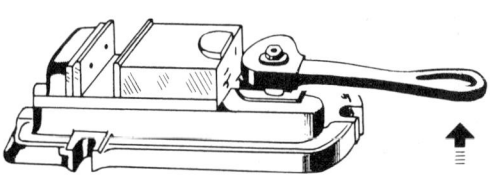 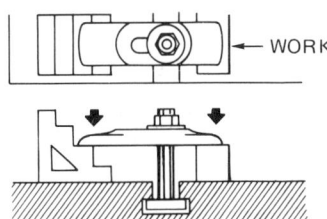

Fig. 11-8

The previous examples are but a few of the many applications of tools which depend on lever principles to control forces and their speeds. In later units the three types of levers are used in combination with other simple machines to produce and control complex and precise movements.

SUMMARY

- A lever is a rigid bar which turns around a fixed point called a fulcrum.
- A lever diagram includes a simple line drawing of the lever; the known and required forces and distances are identified on the diagram.
- The fulcrum of a Class 1 lever is located between the effort and resistance forces.

- The fulcrum of a Class 2 lever is placed at one end and the resistance force is between the fulcrum and the effort force.
- The fulcrum of a Class 3 lever is placed at one end and the effort force is between the fulcrum and the resistance force.
- Regardless of the class of the lever, effort (E) × effort distance (ED) = resistance (R) × resistance distance (RD).
- The weight of the lever may be neglected in problems requiring only an approximate answer. When the lever weight must be computed, the weight is assumed to act at the center of gravity of the lever.
- The mechanical advantage of force of a lever is the ratio of the resistance to the effort force.

$$MA_f = \frac{R}{E}$$

- A lever produces an increase or decrease in the movement of either speed or force depending on the effort (E), resistance (R), effort distance (ED), and resistance distance (RD).
- A lever is balanced when the sum of the moments of force which tend to rotate the lever in one direction is equal to the sum of the moments of force which tend to rotate the lever in the opposite direction.

ASSIGNMENT UNIT 11 SIMPLE MACHINES: LEVERS

■ PRACTICAL PROBLEMS ON LEVERS

Levers of All Classes

For statements 1 to 5, determine which are true (T) and which are false (F).
1. A machine may change the magnitude of a force, its direction, or its speed.
2. A lever diagram of a pair of pliers shows them to be a Class 2 lever.
3. The sum of the (E) moments of any lever is greater than the (R) moments.
4. The fulcrum is a fixed point around which a lever rotates.
5. A lever may be used to obtain the mechanical advantage of both force and speed at the same time.

Select the correct word or phrase to complete statements 6 through 10.
6. By increasing the length of a wrench handle, (less) (more) effort force is required to tighten a nut.
7. A wire-cutting pliers requires less effort to cut a wire when the wire is placed (closer to) (farther away from) the fulcrum.
8. A fluid control valve is regulated by a Class 2 lever. As the distance between the effort force (E) and the fulcrum is decreased, the amount of effort force (E) needed to move the control lever (R) (decreases) (increases).

9. The jaws of a pair of tongs are 3 inches long; the handles are 24 inches long. The mechanical advantage of force is (one-eighth) (eight).

10. Straps and bolts are used to hold parts securely against machine tables to resist the forces exerted by cutting tools. The work is held with (more) (less) force when (ED) is less than (RD).

Class One Levers

1. Name a Class 1 lever that is different from the examples given in this unit.
   a. Diagram the lever.
   b. Identify all parts, forces, and distances.
2. Compute the missing values for A, B, and C in figure 11-9. Disregard the weight of each lever.

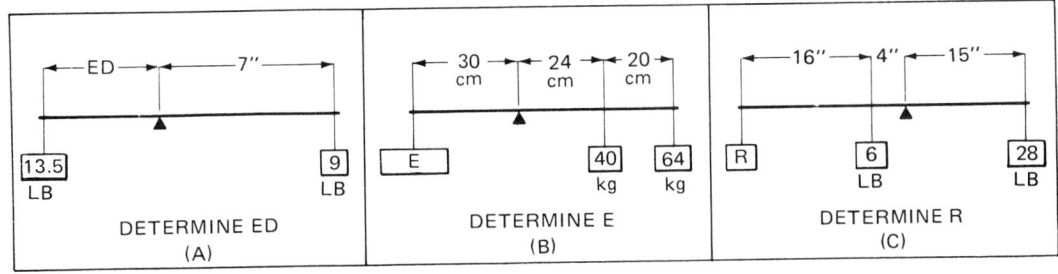

Fig. 11-9

Class Two Levers

1. The center of gravity of a 150-pound load in a wheelbarrow is 16 inches from the fulcrum. What effort (E) is needed to lift the load if the handles are gripped 40 inches from the fulcrum?
2. Name two industrial applications of Class 2 levers as simple machines.
   a. Diagram each lever.
   b. Assign whole number values to the effort and resistance distances and the resistance force of one lever. Solve for the effort force (E).
   c. Assign different whole number values to the effort and resistance forces and the resistance distance. Determine (ED).

Class Three Levers

1. A valve in an industrial production process is controlled by a Class 3 lever. An adjustable weight of two pounds is placed 14 inches from the fulcrum.
   a. Diagram the lever.
   b. Determine the value of the force which this lever arrangement controls (disregard the lever weight).
   c. Compute the (R) force that must be applied five inches from the fulcrum to balance the lever. Disregard the lever weight.

2. Name one other industrial Class 3 lever as a simple machine.
   a. Make a diagram of the lever.
   b. Assign values for (E), (ED), and (RD).
   c. Compute the value of (R).

3. Compute upward forces (F_1 and F_2) required to support the three loads applied on the beam shown in figure 11-10. Assume the fulcrum is on the right end of the beam for the F_2 force.

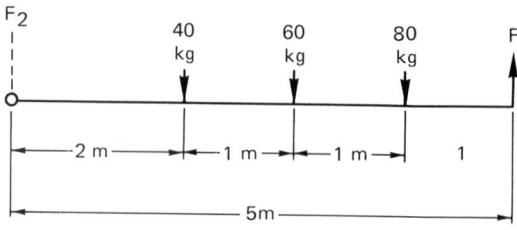

Fig. 11-10

Unit 12 Simple Machines: Inclined Plane and Wedge

The second important simple machine is the inclined plane. The parts and principles of an inclined plane and some practical applications are presented in this unit.

A simple inclined plane is formed by a flat surface at an angle to another surface. The principal advantage of the inclined plane is that it makes it possible to move a load with less effort than is required to lift the same load a vertical distance.

■ PRINCIPLES OF THE INCLINED PLANE

The parts, forces, and dimensions of an inclined plane include: (1) a load (resistance), (2) the force required to move the load (effort), (3) the distance the applied force moves (effort distance), and (4) the vertical height the load is lifted (resistance distance). Although these terms are the same as those used for levers, they are used in a slightly different manner.

An inclined plane is diagramed as a right triangle. The effort force (E) acts parallel to the inclined surface or *slope*. The length of this slope is the effort distance (ED). The object offering the resistance (R) is either pushed or pulled up the slope, figure 12-1.

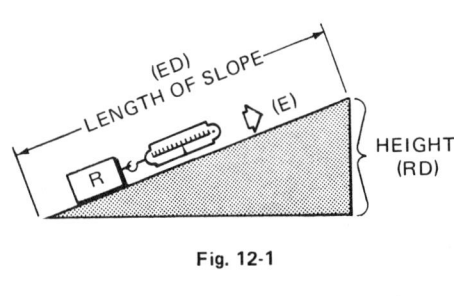

Fig. 12-1

Assume that a 500-pound machine is to be moved from the floor to a platform three feet high, figure 12-2. Using simple machines such as a lever, a set of rollers, and a reinforced inclined plane, this load can be raised and moved where needed.

Figure 12-2 shows a 150-pound effort force pulling the 500-pound weight on a path which is parallel to the 10-foot long slope. (Friction is neglected in this problem).

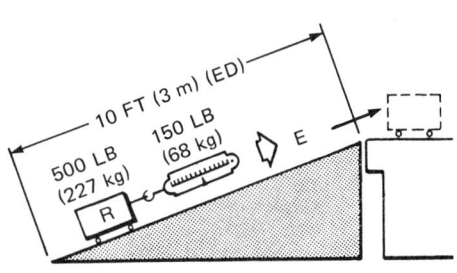

Fig. 12-2

The effort force × the length of the slope is equal to the resistance force × the height the weight is raised. This statement can be expressed as a formula:

$$(E) \times (ED) = (R) \times (RD)$$

■ MECHANICAL ADVANTAGE OF AN INCLINED PLANE

Mechanical Advantage of Force

The 150-pound (68.1-kg) effort force required to lift the 500-pound (227-kg) load indicates that there is some mechanical advantage to be gained by using the inclined plane. The mechanical advantage of force (MAf) is the result of dividing the resistance force (R) by the effort force (E).

$$MA_f = \frac{R}{E}$$

Substituting the values from figure 12-2, the mechanical advantage of force is equal to:

$$MA_f = \frac{500 \text{ lb}}{150 \text{ lb}} = 3\frac{1}{3} \text{ or, } \frac{227 \text{ kg}}{68.1 \text{ kg}} = 3\frac{1}{3}$$

The mechanical advantage can also be found by dividing the distance of the slope (ED) by the vertical rise (RD), or

$$\frac{10 \text{ feet}}{3 \text{ feet}} = 3\frac{1}{3} \text{ or, } \frac{3 \text{m}}{0.9 \text{ m}} = 3\frac{1}{3}$$

Thus, an inclined plane makes it possible to raise a large force using a small force.

Mechanical Advantage of Distance

The gain in force is offset by a loss in distance; that is, the mechanical advantage of distance is equal to the height (RD) divided by the slope (ED).

$$MA_d = \frac{RD}{ED}$$

Substituting the values given in figure 12-2,

$$MA_d = \frac{3 \text{ feet}}{10 \text{ feet}} = \frac{3}{10} \text{ or, } \frac{0.9 \text{ m}}{3 \text{ m}} = \frac{3}{10} = 3:10$$

The mechanical advantage of distance is less than one.

If the slope of this example is increased from 10 feet (3 m) to 15 feet (4.6 m), figure 12-3, the effort required to move the 500-pound load is:

(E) × (15 ft) = (500 lb) × (3 ft)   (E) × (4.6 m) = (227 kg) × (0.9 m)

$$(E) = \frac{1500 \text{ ft-lb}}{15 \text{ lb}} \quad \text{or,} \quad (E) = \frac{204 \text{ m-kg}}{4.6 \text{ m}}$$

(E) = 100 lb   (E) = 44.3 kg

Figure 12-3 shows that the more gradual the slope, the less is the effort required to move a heavy load vertically.

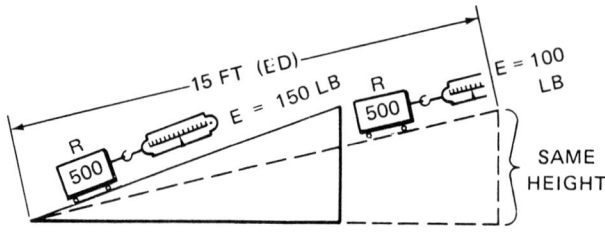

Fig. 12-3 Slope angle affects (E)

■ APPLICATION OF THE INCLINED PLANE TO A WEDGE

There are two basic differences between an inclined plane and a wedge. First, an inclined plane is fixed and does not move. In contrast, the wedge is movable. Second, the effort force moves parallel to the slope of an inclined plane. The effort force for a wedge is applied to the vertical height. The differences between an inclined plane and a wedge are shown in figure 12-4. Note that the resistance (R) moves along the slope of a wedge. The resistance distance (RD) is indicated by the height of the wedge under the resistance.

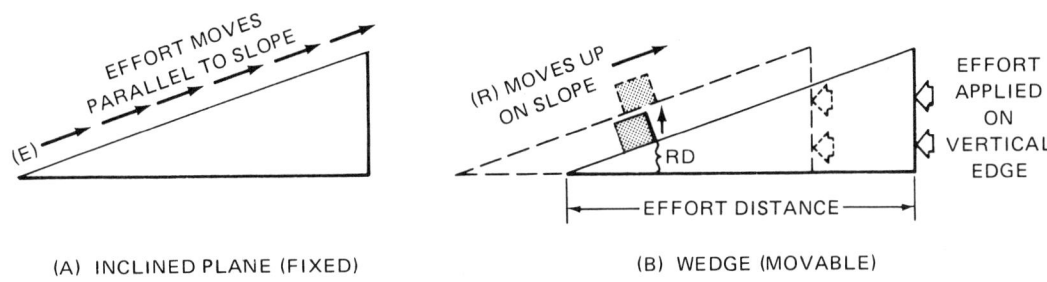

Fig. 12-4

Wedges as Separating Devices

Wedges are used on tools and mechanical devices as separating or holding devices. Wedges are usually illustrated as triangles. However, there are many variations in the actual shapes of wedges. A chisel and an ax are two common applications of the wedge principle to cutting tools.

The cutting lips on a twist drill form a cutting angle. As the drill turns and moves into the work, the material to be drilled is separated in the form of chips. The teeth on a milling cutter, the teeth on a carpenter's saw, a plane blade, and a plumber's pipe cutting rolls all operate on the wedge principle as a separating device, figure 12-5.

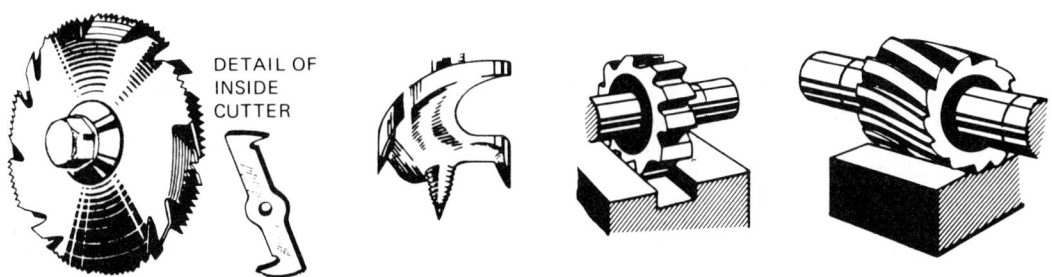

Fig. 12-5 Applications of the wedge as a separating device

Wedges as Holding Devices

The wedge is also used for its holding power, figure 12-6. When a twist drill with a tapered shank is illustrated, it is represented as a wedge, figure 12-6A. Similarly, the bore of the drill press spindle in which the drill is placed is wedge-shaped. Another application of the wedge as a holding device is used in building construction as shown in figure 12-6B. A pair of wedges raises the I-beam and holds it securely in position.

Tool manufacturers have experimented for many years to determine the best angles for holding devices. The holding ability of a wedge decreases as its angle increases beyond the best angle.

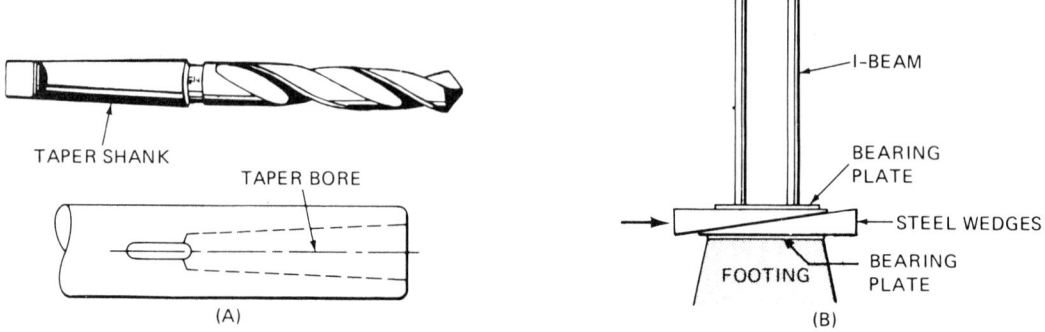

Fig. 12-6 Applications of the wedge as a holding device

■ MECHANICAL ADVANTAGE OF WEDGES

The mechanical advantage of a wedge is found by using the same formulas used previously for the inclined plane. In wedge problems, care must be taken to assign the correct values for (E), (ED), (R), and (RD).

Cams are mechanical devices used on machines to impart motion. Cams operate on the wedge principle. A follower bearing on the wedge surface produces motion in a second mechanism.

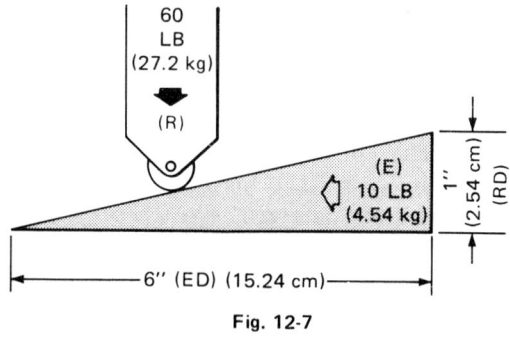

Fig. 12-7

A diagram of a cam, figure 12-7, shows that it has a mechanical advantage of force of six and a mechanical advantage of distance (or speed) of one-sixth.

Mechanical advantage of force:

$$MA_f = \frac{R}{E} = \frac{60 \text{ lb}}{10 \text{ lb}} = 6 \text{ or, } \frac{27.2 \text{ kg}}{4.54 \text{ kg}} = 6$$

Mechanical advantage of distance:

$$MA_d = \frac{RD}{ED} = \frac{1''}{6''} = \frac{1}{6} \text{ or, } \frac{2.54 \text{ cm}}{15.24 \text{ cm}} = \frac{1}{6}$$

It is apparent that the mechanical advantage of a wedge is influenced by the angle. As the angle increases, the MA_f decreases.

This unit has covered the application of the principles of the inclined plane to wedges. Another unit will cover the application of these principles to screw threads.

SUMMARY

- An inclined plane is used when a large resistance is to be moved vertically by a comparatively small effort force.
- The effort force used with an inclined plane varies with the angle. The larger the angle of the inclined plane, the greater is the effort force required to produce work.
- The mechanical advantage of force of an inclined plane = resistance force divided by effort force (R/E).
- The mechanical advantage of distance or speed of an inclined plane = resistance distance divided by effort distance (RD/ED).
- The wedge is a practical application of the inclined plane to separating and holding devices.
- A wedge is diagramed as a right triangle with the base representing the effort distance (ED); the height is the effort (E); the resistance moving up the slope is (R); and the vertical distance the resistance moves is (RD).
- For any wedge, effort (E) × effort distance (ED) = (R) × (RD).
- The mechanical advantage of a wedge is computed using the same formulas as for the inclined plane.

ASSIGNMENT UNIT 12 SIMPLE MACHINES: INCLINED PLANE AND WEDGE

■ PRACTICAL PROBLEMS WITH THE INCLINED PLANE AND WEDGE

The Inclined Plane

Select the correct word or phrase to complete statements 1 to 6.

1. The vertical distance a load is moved up an inclined plane is the (effort distance) (resistance distance).
2. The effort force on an inclined plane acts parallel to the (slope) (base) (height).
3. Inclined planes are used because there is a (gain) (loss) in the mechanical advantage of force.
4. The MA_f and MA_d for an inclined plane are (the same) (different).

108 Mechanics, Machines, and Wave Motion

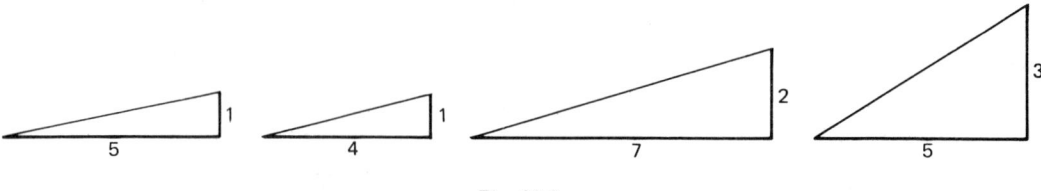

Fig. 12-8

5. The MA_f in each of the four diagrams in figure 12-8 (is equal) (varies).
6. As the (RD) on an inclined plane is increased, the MA_f (increases) (decreases).
7. For parts A, B, C, D, and E in the following table, compute the forces or distances required to do the work specified using an inclined plane. (Disregard friction and round off numbers to the nearest whole number.)

	Weight of Castings (R)	Height to be Raised (RD)	Effort Force (E)	Slope Length (ED)
A		4 ft	100 lb	8 ft
B	200 lb		50 lb	16 ft
C	200 lb	4 ft		4 ft
D		4 ft	133 lb	6 ft
E	200 lb	4 ft	57 lb	

8. Make a graph of (ED) and (E) using the data in the table from Problem 7. Use graph paper with 1/4-inch squares. For (ED) use a horizontal scale of 1 inch = 2 ft, and for (E) a vertical scale of 1 inch = 50 lb. Indicate the scales on the graph.
9. Determine from the graph prepared for Problem 8 what (E) is required to move the 200-pound castings vertically four feet for slope lengths of 10 ft, 11 ft, and 12 ft.
10. Check the accuracy of the graph readings by computation. Show all mathematical processes.
11. Convert the customary foot and pound units on the graph for Problem 8 to equivalent meter and kilogram units. Add the metric set of values to the graph.
12. Read the metric values on the graph for problem 11 to determine the force required to move 90.8-kg castings vertically 1.2 m on (a) 2.1 m and (b) 3.3 m length inches.

The Wedge

Select the correct word or phrase to complete statements 1 to 5.
1. A wedge differs from an inclined plane in that it is (fixed) (movable).
2. The effort force on a wedge is applied along the (vertical edge) (slope) (base).

3. The holding power of a wedge depends on the (angle of the wedge) (length of the slope).

4. The (slope) (height) (base) represents the resistance distance of a wedge.

5. The mechanical advantage of distance for a wedge is always (less than one) (more than one).

Add a word or phrase to complete statements 6 to 9.

6. The wedge may be used as a _____ or a _____ device.

7. The holding power of a wedge _____ as the inclined angle of the wedge increases beyond the best angle.

8. The mechanical advantage of force of a wedge equals _____.

9. The base of a wedge represents the _____.

10. List three industrial applications of the wedge as a separating device and two other applications of a wedge as a holding device.

11. A wedge-shaped cam and a follower are used in a manufacturing process. The input driving force (E) of 100 pounds (45.4 kg) moves at a constant distance (ED) of one inch (2.54 cm). The follower, representing (R), moves the distances given in the table. Find (R) for cams A, B, C, and D.

	Wedge Angle of Cam	Driving Force (E)	Effort Distance (ED)	Resistance Force (R)	Follower Movement (RD)
A	15°	100 lb	1 in		1/3 in
B	30°	100 lb	1 in		2/3 in
C	45°	45.4 kg	2.54 cm		2.54 cm
D	60°	45.4 kg	2.54 cm		3.4 cm

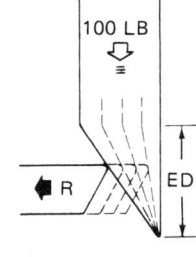

Fig. 12-9

12. Graph the relationship between the wedge angle of the cam and the resistance force (R) which can be overcome. Use a vertical scale of 1 inch = 15° and a horizontal scale of 1 inch = 50 pounds and 22.7 kg on graph paper with 1/4-inch squares. Indicate the scales and values on the graph.

13. What kind of application of the wedge is a cam and follower?

14. Interpret the graph prepared for Problem 12 in terms of the effect on (R) of increasing the wedge angle.

Unit 13 Simple Machines: The Wheel and Axle

The wheel and axle is a simple machine consisting of a large handle or circular part (the wheel) which is rigidly secured to a smaller circular part (the axle). The wheel and axle serves one of two basic functions: (1) to transmit force or (2) to produce a change in speed.

■ THE WHEEL AND AXLE DIAGRAM AND PARTS

A common example of the wheel and axle is a set of pulleys. The larger pulley is the wheel and the smaller pulley is the axle. Figure 13-1 shows that the radius of the axle is the resistance distance (RD) and the pull of the belt on the axle is the resistance (R).

The moments of force tend to revolve about the fulcrum (the fulcrum is at the center axis of the wheel and axle). The effort distance (ED) is the radius of the wheel and the effort force is represented by (E). Part of the diagram showing these forces and distances represents a Class I lever. Recall that the wheel and axle is a practical application of the lever. The symbols used on the diagram of the wheel and axle are the same as those applied to the lever.

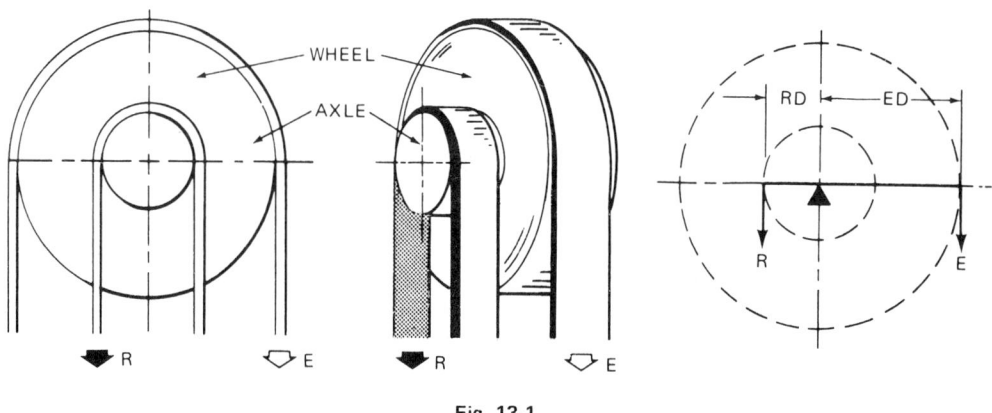

Fig. 13-1

■ MOMENTS OF FORCE OF THE WHEEL AND AXLE

The moments of force of the wheel and axle are represented by the effort force and effort distance of the wheel and the resistance force and resistance distance of the axle. In the perfect wheel and axle, these moments of force are equal: (E) × (ED) = (R) × (RD).

Simple Machines: The Wheel and Axle

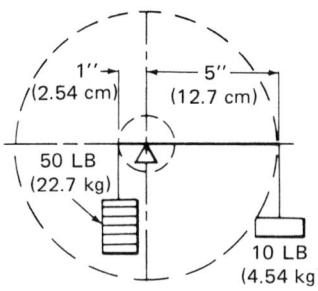

Fig. 13-2

Figure 13-2 shows an effort force of ten pounds applied to a wheel with a radius of five inches. The resistance force acting on an axle with a radius of one inch (the force the effort force is to overcome) is found by the moment formula: (E) × (ED) = (R) × (RD). Substituting the values shown in figure 13-2, the result is:

(10 lb) × (5 in) = (R) × (1 in)   (4.54 kg) × (12.7 cm) = (R) × (2.54 cm)

$$\frac{50 \text{ pound-inches}}{1 \text{ inch}} = (R) \qquad \text{or,} \qquad \frac{57.66 \text{ kg-cm}}{2.54 \text{ cm}} = (R)$$

50 pounds = (R)   22.7 kg = (R)

Thus, a wheel and axle combination makes it possible to use a smaller force to produce or control a larger force.

■ MECHANICAL ADVANTAGE OF FORCE

The amount of effort force needed to move a resistance depends upon the distance of both forces from the fulcrum. As the effort distance increases, the same effort force may be used to overcome a larger resistance force. In other words, the mechanical advantage of force is increased Similarly, the 22.7 kg ÷ 4.54 kg = MA$_f$ of 5.

$$MA_f = \frac{R}{E}$$

If an effort force of 10 pounds on the wheel overcomes a resistance force of 50 pounds on the axle, the MA$_f$ = 50/10 = 5.

The mechanical advantage of force for a wheel and axle differs from that of the lever. That is, the wheel and axle rotates and the mechanical advantage is constant. By contrast, the lever must be raised and lowered every time the force is applied.

■ CHANGING SPEEDS USING THE WHEEL AND AXLE

A point on the outside rim or *periphery* of the wheel moves a greater distance than a corresponding point on the axle. This is due to the fact that the circumference of the wheel is greater than the circumference of the axle, figure 13-3.

112 Mechanics, Machines, and Wave Motion

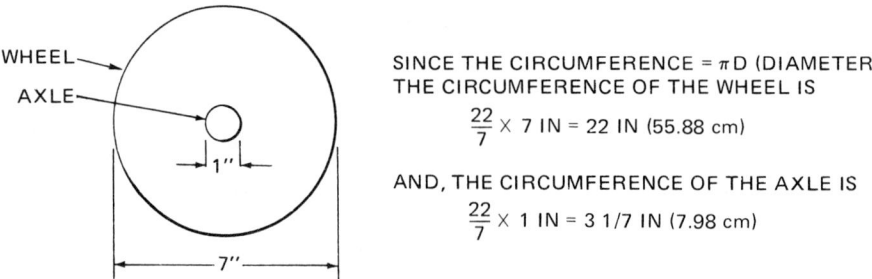

SINCE THE CIRCUMFERENCE = πD (DIAMETER):
THE CIRCUMFERENCE OF THE WHEEL IS

$\frac{22}{7}$ × 7 IN = 22 IN (55.88 cm)

AND, THE CIRCUMFERENCE OF THE AXLE IS

$\frac{22}{7}$ × 1 IN = 3 1/7 IN (7.98 cm)

Fig. 13-3

For the dimensions shown in figure 13-3, the circumference of the wheel is seven times greater than the circumference of the smaller pulley or axle. If the wheel is driving the axle, there is a loss in the mechanical advantage of speed. This loss is due to the fact that the driving belt travels seven times faster and farther than the driven belt on the axle.

Figures 13-4A and 13-4B show that any gain in the mechanical advantage of force is offset by a loss in the distance traveled or a loss of speed. Whenever the wheel and axle is used as a simple machine, only the force *or* the speed may be increased. The mechanical advantage of speed is equal to the resistance distance divided by the effort distance.

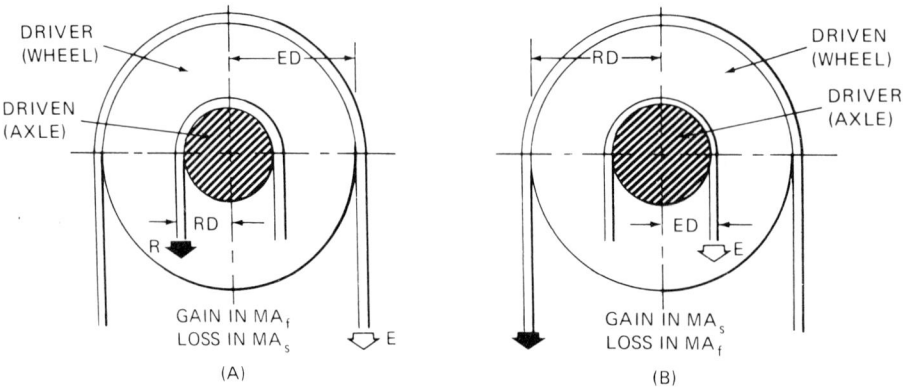

Fig. 13-4

■ INDUSTRIAL APPLICATIONS OF THE WHEEL AND AXLE

The applications of the wheel and axle are designed to achieve either of two goals: (1) to obtain a gain in force, or (2) to control speed, figure 13-5. The first type of application includes cranks and handwheels having large radii which turn smaller axles to raise, lower, or adjust machine tables, columns, or other mechanical devices.

Simple Machines: The Wheel and Axle 113

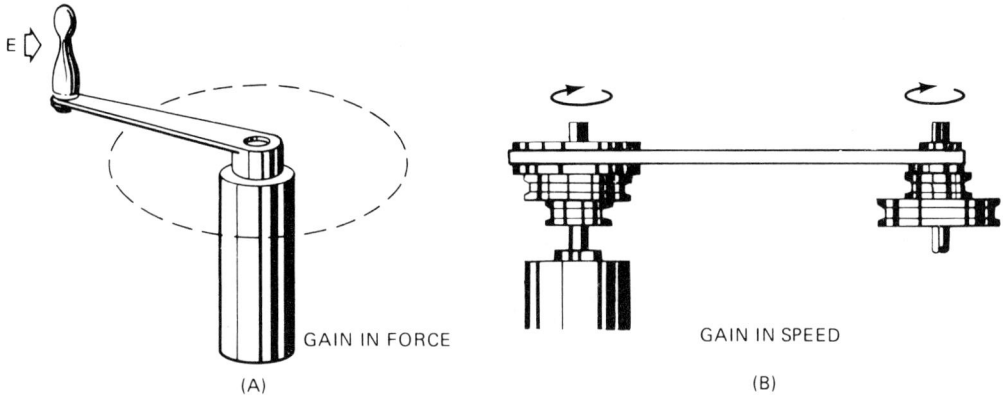

Fig. 13-5 Wheel and axle applications

Step-cone pulleys are an excellent example of the wheel and axle principle as it is used to control speed. Figure 13-5B shows the axle of a motor to be the driver with the largest pulley driving a smaller pulley. Such a combination produces a gain in speed on the machine spindle. Many other applications of the lever and wheel and axle principles are shown by simple gears, flywheels, and other rotating parts which gain either force or speed.

SUMMARY

- In industry, the wheel and axle principle is used to transmit force or to change speed.
- The diagram of a wheel and axle is similar to a lever diagram and includes the same terms.
- For a wheel and axle, the effort moments equal the resistance moments.

    Effort (E) × Effort Distance (ED) =
    Resistance (R) × Resistance Distance (RD)

- The mechanical advantage of force = $\dfrac{R}{E}$.

- The mechanical advantage of speed = $\dfrac{RD}{ED}$.

- When there is a gain in force, there is loss in speed. Conversely, a gain in speed is offset by a loss in force.
- On simple machines, it is impossible to gain both force and speed at the same time.

ASSIGNMENT UNIT 13 SIMPLE MACHINES: THE WHEEL AND AXLE

■ PRACTICAL PROBLEMS WITH THE WHEEL AND AXLE

Moments of Force on the Wheel and Axle

Select the correct word or phrase to complete statements 1 to 6.

1. The sum of the effort moments and the sum of the resistance moments of a wheel and axle are (different) (equal).

2. An effort force, acting at a larger radius, will move a (heavier) (lighter) resistance force.

3. The moment formula for the wheel and axle is (E × R = ED × RD) (E × ED = R × RD).

4. The wheel and axle produce (both a gain in force and a gain in speed) (either a gain in force or a gain in speed).

5. The mechanical advantage of force is $\left(\dfrac{R}{E}\right)\left(\dfrac{RD}{ED}\right)$.

6. The mechanical advantage of speed is $\left(\dfrac{R}{E}\right)\left(\dfrac{RD}{ED}\right)$.

7. Compute the effort force, the resistance force, or the resistance distance for the crank arm combinations indicated in the table at A, B, C, D, and E.

	Wheel and Axle Combinations			
	Resistance	Effort	(RD)	(ED)
A	75 lb	7.5 lb	1 in	
B	500 lb	125 lb		12 in
C	1250 lb		6 in	12 in
D		450 kg	24 cm	30 cm
E	1020 kg	814 kg	30 cm	

8. Determine the mechanical advantage of force for the wheel and axle combinations given in the table in Question 7. Use the computed values for the missing values.

Controlling Speed with the Wheel and Axle

Add a single word to complete statements 1 to 4.

1. A point on the rim of a wheel moves _____ than the corresponding point on the axis.

2. When the effort force is applied on the axle, the _____ force moves faster.

3. A gain in the speed of a wheel and axle combination is offset by the _____ _____ in the force.

4. Speed is developed on a wheel and axle when the effort is applied to the _____ _____.

5. Compute the MA_S for the four pulley combinations given in the table for each of the three different sizes of pulleys shown.

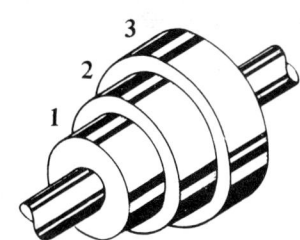

	Radius of Axle	Radius of Pulleys		
		1	2	3
A	1/2"	2"	3"	4"
B	1"	2"	3"	4"
C	50.8 mm	50.8 mm	76.2 mm	101.6 mm
D	50.8 mm	50.8 mm	152.4 mm	203.2 mm

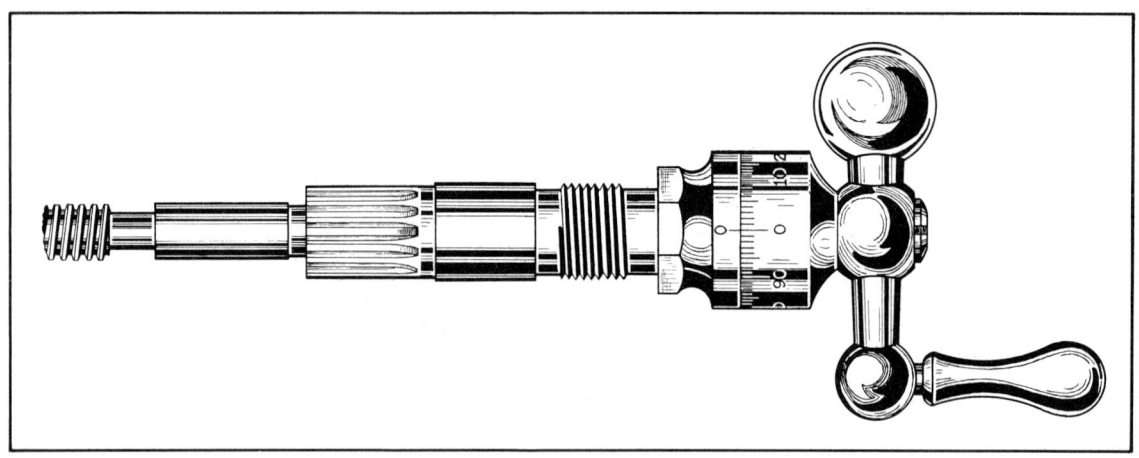

Threaded feed screw rotary motion converted into calibrated feed measurements

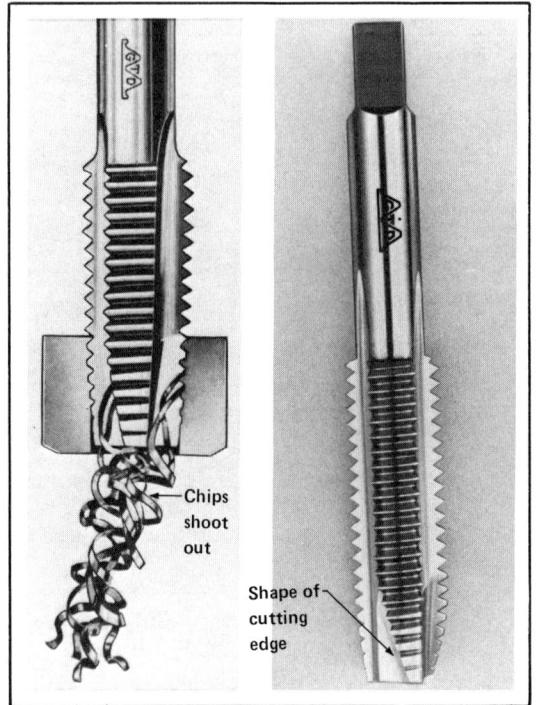

Cutting internal threads with a tap (Courtesy Greenfield Tap and Die Division, TRW Inc.)

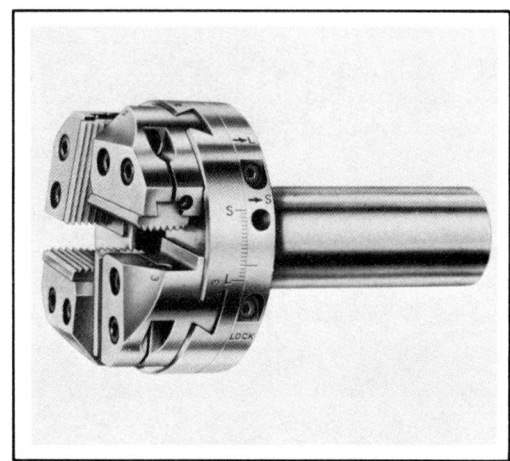

Producing external threads with a self-opening die and thread chasers (Courtesy Greenfield Geometric Tool Division, TRW Inc.)

Unit 14 Simple Machines: The Screw Thread

A screw thread is one of the most practical applications of the principles governing the inclined plane, the wedge, and the wheel and axle. A *screw thread* is a ridge of uniform cross section formed around a cylinder or cone. There is a constant distance between each thread.

■ SCREW THREAD USES AND TERMS

Screw threads are used (1) to transmit motion, (2) to apply tremendous forces with comparatively little effort, (3) to hold parts together, and (4) to obtain measurements.

Screw threads are specified according to their size, the number of threads per inch, and the form of the thread. Most screw threads are said to have a definite number of threads per inch. For example, in a single screw thread with ten threads per inch, the distance between two successive threads is one-tenth of an inch. This distance is called the *pitch*, figure 14-1. The pitch is important when computing the force a screw exerts, determining the mechanical advantage of a screw thread, and measuring.

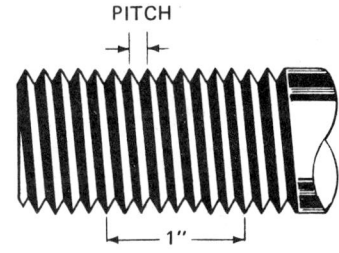

Fig. 14-1

$$\text{Pitch (P)} = \frac{1}{\text{number of threads per inch}}$$

■ TRANSMITTING MOTION WITH THE SCREW THREAD

Screw threads are generated around a circular part. As a screw thread is turned, it transforms rotary motion to rectilinear (straight line) motion. The distance traveled by the part in contact with the screw thread depends upon the pitch of the screw and the amount the screw is turned.

A screw thread with a pitch of 1/10 inch (2.5 mm), figure 14-2, produces a rectilinear motion of 0.100 inch (2.5 mm) for each complete revolution. The rectilinear

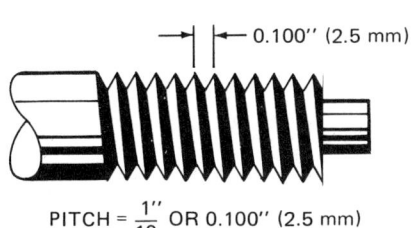
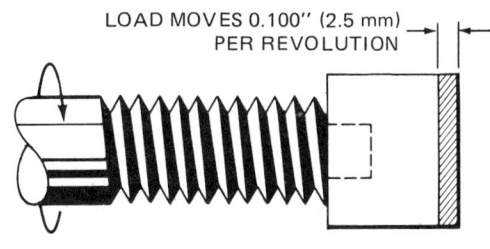

Fig. 14-2

motion is governed by the circular motion of the screw thread. This characteristic of screw threads is applied to lead and feed screws. When such screws are turned, they move a table or column in a vertical or horizontal direction.

■ SCREW THREADS USED IN MAKING ADJUSTMENTS AND MEASUREMENTS

One of the most common applications of the screw thread is to make fine adjustments such as in measuring devices. The legs on a drawing compass, a divider, a caliper, and the beam on a vernier caliper are all adjusted by a screw thread. The tables or elevating columns on precision machines are moved specified distances by attaching a graduated collar to a screw thread.

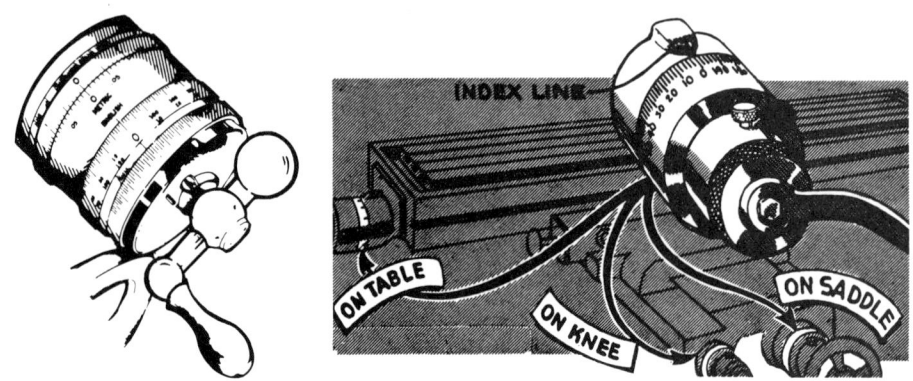

Fig. 14-3 Machine applications of screw threads

For example, if a lead screw on a machine has ten threads per inch, the pitch of the screw is one-tenth of an inch. As the screw advances one complete turn, the table to which the screw is attached advances this same distance (0.100 in). If a circular collar having 100 graduations around its periphery is fastened to the lead screw, the distance the table moves from one graduation on the collar to the next is 1/100 of 0.100 inch, or one-thousandth of an inch (0.001 in) or twenty-five thousandths of a millimeter (0.025 mm).

This combination of the pitch of a screw thread and a graduated collar is commonly used on measuring instruments and mechanical devices. As a result, accurate measurements can be made easily.

At times a thread must be moved a required distance with fewer turns. When a screw thread moves twice the distance for each complete turn as it does for a single thread, it is known as a *double pitch* screw thread. Similarly, screws which advance three times as far in one turn have a *triple pitch* screw thread; a distance of four times as far indicates a *quadruple pitch* screw thread, and so forth.

Double, triple, and quadruple pitch threads are classed as *multiple threads*. The term *lead* is used with multiple threads to indicate how far the screw moves for each complete turn. Lead should not be confused with pitch. Pitch always means the distance between one thread and a corresponding point on the next thread. The *lead* of a multiple thread is the distance the screw moves with each complete turn.

■ SCREW THREADS USED AS FASTENING DEVICES

Screw threads also act as wedges to give a holding action. Once screw threads lift a heavy load, they must be capable of holding the load in position. The screw on a vise has three functions: it is used to adjust the vise, it applies force to hold a workpiece, and the threads act as wedges to prevent the part from moving.

Machine and automotive bolts are common uses of the principle of the thread as a fastening device to hold parts together securely. In this type of application, the pitch of the screw threads is useful in determining the holding power.

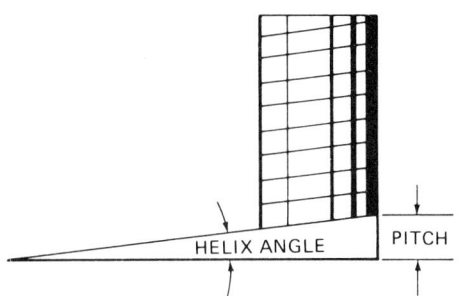

Fig. 14-4 Inclined plane applied to thread

■ TRANSMITTING FORCE WITH SCREW THREADS

The screw thread is used to transmit great forces by either pushing or pulling. In this respect, the mechanical advantages of the inclined plane and the lever are combined to create a tremendous force.

One of the common applications of a screw thread is to a jack as shown in figure 14-5. There are several mechanical advantages of force which are provided by this device: (1) there is the mechanical advantage of the inclined plane which represents the *helix angle* of the thread, (2) there is the mechanical advantage of the lever which turns the screw, and (3) the combination of the screw thread and the lever result in a wheel and axle which provides another mechanical advantage.

The jack in figure 14-5 is raising an I-beam. This beam, in turn, supports another load. The screw has four threads per inch, or a pitch of 0.250 inch (6.35 mm). One complete turn of the 21-inch (0.53-m) crank handle moves the I-beam 0.250 inch (6.35 mm or 0.006 35 m). The effort force moves a distance equal to the circumference of a circle with a radius of 21 inches. The circumference is equal to pi (π) $\times$ diameter. Thus, the circumference = 22/7 $\times$ 42 inches = 132 inches (3.35 m).

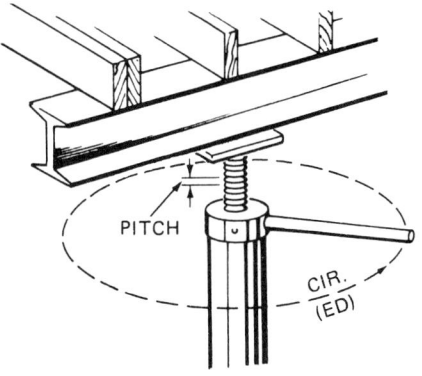

Fig. 14-5

120 Mechanics, Machines, and Wave Motion

The mechanical advantage is given by ED/RD.

$$\text{MA} = \frac{ED}{RD} \quad \text{or,} \quad \text{MA} = \frac{3.35}{0.00635}$$

$$= 132 \div \frac{1}{4} \qquad = 528$$

$$= 132 \times \frac{4}{1}$$

$$= 528$$

A mechanical advantage of 528 means that an effort force of 10 pounds (4.5 kg) is capable of moving a resistance force of 5280 pounds (2377 kg). (Friction losses are neglected in this example.) If the effort distance increases, the same effort force will lift an even greater load.

If the length of the lever handle is increased from 21 inches to 42 inches (by adding a length of pipe, for example), the ED = (pi) × (diameter) = 22/7 × 84 inches = 264 inches. The mechanical advantage of this combination can be calculated by substituting (ED) = 264 and (RD) = 1/4 in the expression MA = ED/RD.

$$\text{MA} = 264 \div \frac{1}{4}$$

$$= \frac{264}{1} \times \frac{4}{1}$$

$$= 1056$$

In other words, a ten-pound (4.5-kg) effort force applied to the 42-inch (1.07-m) handle will raise a load of (1056) × (10) or 10 560 pounds (4794 kg).

It is not possible to increase the effort distance indefinitely because of the physical limitations of materials. However, this simple example shows the way in which large resistance forces can be overcome by smaller effort forces. In actual practice, the mechanical advantage of the screw thread is somewhat less than the computed values because it is necessary to overcome friction losses.

The thread angle should be as small as possible and still have the required strength. The greater the thread angle, the greater is the force which tends to break the part into which the screw thread fits.

As the number of threads per inch increases, the helix angle decreases, figure 14-6. The *helix angle* corresponds to the angle of an inclined plane. The pitch of the thread

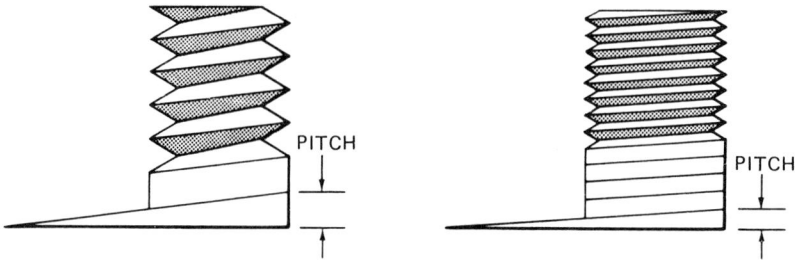

Fig. 14-6 Helix angle affects pitch

is the same as the RD and the length of the slope is the ED. Since the greatest mechanical advantage is obtained when a small (E) force moves a large (R) force, a decrease in the helix angle means that the holding power of the thread is increased.

This holding power of the screw thread should not be confused with the force that a screw thread can exert. As the thread becomes finer (the pitch decreases), the thread cuts become shallower. As a result, the application of a great force to a screw with a fine pitch may strip the screw thread.

■ CHARACTERISTICS OF COMMON THREAD FORMS

The most commonly used thread forms are the Unified and the American National forms. The only difference between these two forms is in the shape of the top (crest) and the bottom (root) of the thread. Both the crest and root of the American National thread form are flat. In contrast, the crest of the Unified thread form may be either flat or rounded and the root is always rounded. The characteristics of the two thread forms are shown in figure 14-7.

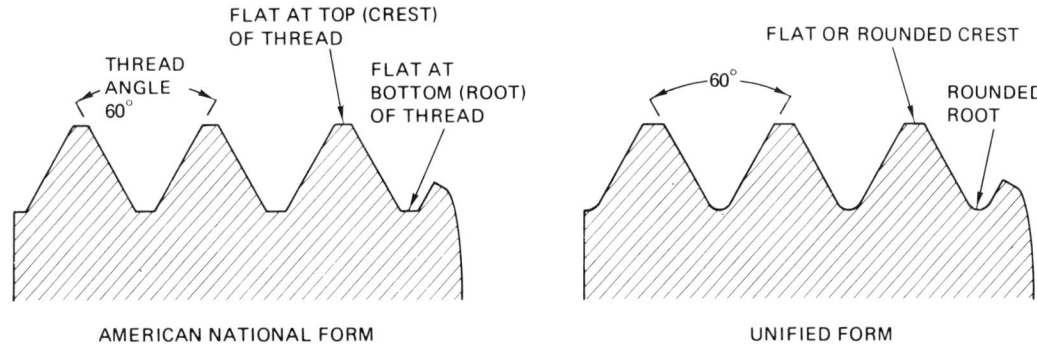

Fig. 14-7 Characteristics of thread forms

Common Thread Forms

Figure 14-8 shows the pictorial representations of four common thread forms using the 60° included angle. The selection of the thread form to be used depends

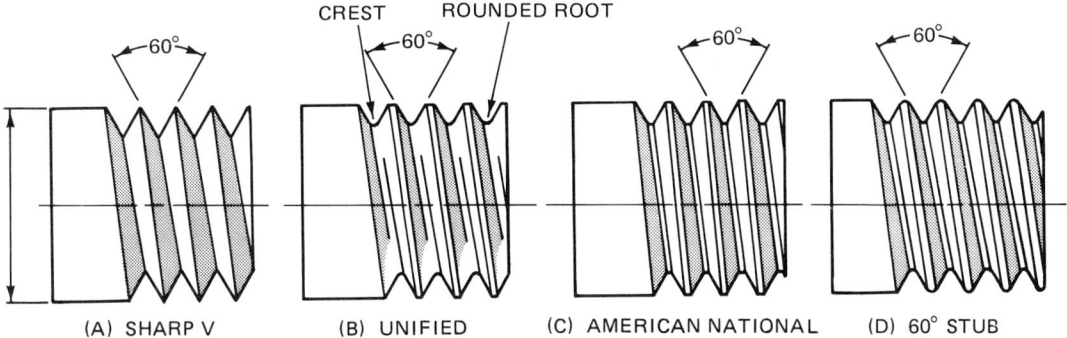

Fig. 14-8 Common thread forms

upon the application. For example, the 60° *stub thread* is used where the Unified and American National forms may have too much depth for a particular part.

A square thread form or a modified form is used when great power or force is needed. The square form, figure 14-9A, is called a *square thread.* The more popular modified thread form has an included angle of 29° and is known as the *acme thread,* figure 14-9B. The *buttress thread,* figure 14-9C, is used where power is to be transmitted in one direction. The *knuckle thread,* figure 14-9D, is often used when a thread is molded (as in ceramic parts), or is rolled into thin metal parts (such as lamp sockets).

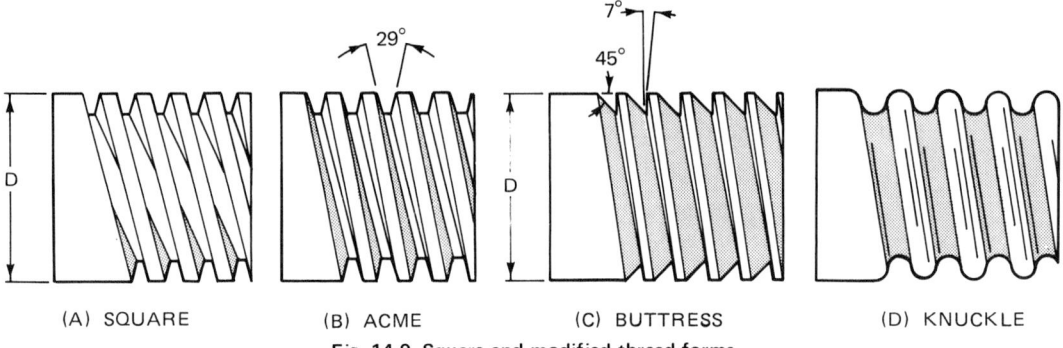

(A) SQUARE   (B) ACME   (C) BUTTRESS   (D) KNUCKLE

Fig. 14-9 Square and modified thread forms

■ OTHER PRACTICAL APPLICATIONS OF SCREW THREADS

A nut on a saw arbor or the spindle of a grinder are typical applications of the screw thread as a holding device. The micrometer combines the principles of the screw thread and the graduated collar to make measurements to accuracies of (0.001 in) and (0.0001 in). The screw on a forming press has two functions in that it changes the rotary motion of a handwheel to a rectilinear motion and develops a large force against the part to be formed. The list of applications of screw threads is an endless one.

Screw threads of different forms are available with different types of heads. The square head machine screw, the hexagon head machine screw, and the Allen head setscrew are common types of screws used as fasteners.

SUMMARY

- A screw thread is a combination of two simple machines: an inclined plane which forms the helix angle and the wedge as applied to the teeth.
- Screw threads are used to:
  — change rotary motion to rectilinear or straight line motion,
  — increase force,
  — fasten parts together, and
  — make adjustments and take measurements.

- A third simple machine, the wheel and axle, is sometimes combined with a screw thread to produce a greater mechanical advantage of force.
- The pitch of a single thread screw is the distance from one point on one tooth to a corresponding point on the next tooth.
- The lead is the distance that a thread moves with each complete revolution.
- The pitch formula is:

$$P = \frac{1}{\text{number of threads per inch}}$$

- The helix angle of a screw thread determines the holding power of the thread if the physical properties of the material are not exceeded.

ASSIGNMENT UNIT 14 SIMPLE MACHINES: THE SCREW THREAD

■ PRACTICAL PROBLEMS WITH FOUR MAJOR APPLICATIONS OF SCREW THREADS

For statements 1 to 7, determine which are true (T) and which are false (F).

1. The pitch of a thread refers to the shape of the thread.
2. For the same diameter, a screw with a coarser thread can exert a greater (E) force than a screw with a fine pitch thread.
3. A screw thread is an application of the inclined plane and wedge.
4. A fine pitch screw thread travels a greater distance each revolution than does a coarse pitch screw thread.
5. The screws on an engine head serve as fasteners to hold the head securely to the engine block.
6. The threads on a jack used in building construction change a rotary motion of the screw to a straight line motion to move heavy loads.
7. The threads on a measuring instrument are used to transmit tremendous forces.
8. Prove or disprove by scientific reasoning the statement that the finer the pitch of a screw thread, the greater is the force that may be applied.
9. State three industrial applications (different from those given in this unit) of screws for each of the following:
    a. as measuring devices,
    b. to increase force,
    c. to change rotary motion to straight line motion, and
    d. as fasteners.

10. Determine the load which can be raised with the elevating screw combinations given in the table at A, B, C, D, E and F. (Disregard friction losses.)

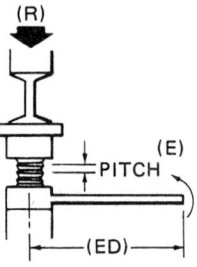

Fig. 14-10

	A	B	C	D	E	F
Threads per inch (TPI) or millimeter pitch (mm)	10 TPI	8 TPI	4 TPI	12 mm	6.35 mm	12.70 mm
Overall radius of lever (ED)	28"	28"	28"	72 cm	1.07 m	1.07 m
Effort force (E)	40 lb	40 lb	40 lb	18 kg	40 kg	80 kg
Load (R)						

11. Graph the approximate relationship between the pitch and the load for combinations A, B, and C (in problem 10) where the values of (ED) and (E) remain constant. Represent the threads per inch on a horizontal scale of 1 inch = 2 threads; the vertical scale is 1 inch = 12 000 pounds. Round off the computed load (R) to the nearest 1000 pounds.

12. Determine the approximate loads that screws with 3, 5, and 9 threads per inch can raise. Assume that (ED) is 28 inches and (E) is 40 pounds.

13. Calculate the load that can be raised by three screw mechanisms which have metric pitches of (a) 8 mm, (b) 5 mm, and (c) 3 mm. The (ED) is 72 cm and the (E) force on the lever is 18 kg in each case.

Single gear sets and compound gear trains applied to the driving mechanism of a machine tool spindle (Courtesy Lodge & Shipley Co.)

Unit 15 Simple and Compound Gear Trains

Machines and other mechanical devices often depend upon gears and gear systems for their operation. Gears have four functions: (1) they transmit positive motion; (2) they are used to drive a second part or mechanism in either the same or the opposite direction to that of the motor or other driving force; (3) they are used to drive or control a second gear or part at the same speed, or at a slower or faster speed than the driver; and (4) they can be used to increase or decrease force. There are many types, shapes, and sizes of gears; however, the principles governing each of the four major gear functions apply to all gears.

■ SIMPLE GEAR TRAINS

Motion can be transmitted from one smooth-surfaced cylinder to another by friction. If teeth are added around each cylinder, gears are formed. As one tooth in one gear drives against a tooth in a mating gear, there is positive contact and motion is transmitted.

When two gears are in mesh, the gears are known as a *pair of gears*.

As the number of gears is increased to three or more, the combination is called a *simple gear train*, figure 15-1. Simple gear trains can perform the same four basic functions listed previously. The following sections of this unit present each function as it is carried out by a simple gear train and then by a compound gear train.

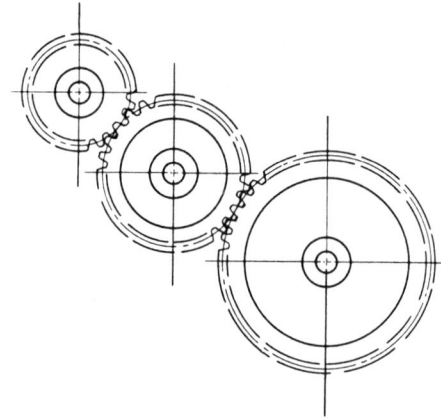

Fig. 15-1 Representation of a simple gear train

■ GEARS TRANSMIT POSITIVE ROTARY MOTION

Motion can be transmitted by pulleys or shafts which either revolve against each other or are driven by belting. In these situations, there is a considerable loss in output due to slippage.

As the teeth added to circular parts are brought into contact, they will *mesh* if they have the same shape, depth, and thickness. The meshing action of one tooth on a driver gear against another tooth on a driven gear produces positive rotary motion without slippage. This feature is desirable when constant motion is required.

■ GEARS CHANGE THE DIRECTION OF ROTATION

Gears in mesh with each other revolve in opposite directions. Note in figure 15-2 that as gear (A) revolves clockwise, it drives a second gear (B) counterclockwise. The gear that is rotated by the source of power is called the *driver* gear. The last gear on a train is known as the *driven* gear. If a third gear is added between the pair of gears, the driver and driven gears rotate in the same direction.

The gears placed between the driver and driven gears are called *intermediate gears*. These gears control the direction of the driven gear. Intermediate gears also make it possible to compensate for changes in overall dimensions when gears of different sizes are used. The addition of a second intermediate gear causes the rotation of the driven gear to be in the direction opposite to that of the driver.

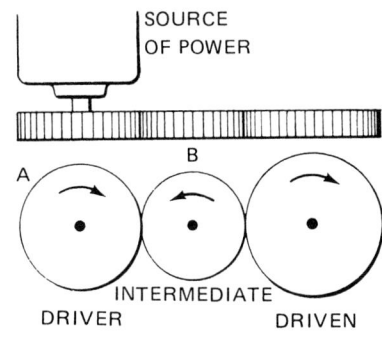

Fig. 15-2

Regardless of the number of intermediate gears in a simple gear train, the ratio between the driver gear and the driven gear remains the same.

- The driver and driven gears rotate in the same direction when there is an odd number of gears in a simple gear train.
- The gears rotate in opposite directions when there is an even number of gears in the gear train.

■ GEARS CHANGE SPEEDS

If two gears have the same size and number of teeth and are in mesh, they rotate at the same speed. Any change in the number of teeth in one gear affects the speed of the other gear. In this manner, the driven gear can be turned a desired number of revolutions.

Figure 15-3 shows three pairs of gears. For the gears at A, the ratio is 1:1 since both gears have the same number of teeth (64). Both the driver gear and the driven

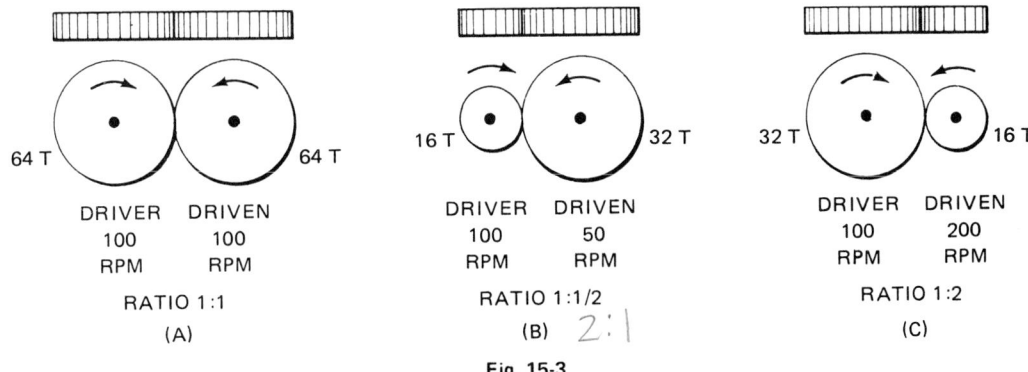

Fig. 15-3

gear turn at 100 revolutions per minute (rpm). In figure 15-3B, a 16-tooth gear drives a 32-tooth gear. The driver gear must revolve twice to turn the driven gear one complete revolution. As a result, the speed of the driven gear is half the original value or 50 revolutions per minute. The gears in figure 15-3C consist of a 32-tooth driver gear and a 16-tooth driven gear. In one complete revolution of the driver, the driven gear makes two complete revolutions. Thus, a driver gear speed of 100 rpm results in a driven gear speed of 200 rpm.

Ratio of Speed

The ratio of speed is the speed of the driver gear divided by the speed of the driven gear. For figure 15-3A, the speed ratio is 1:1. This ratio is obtained by dividing the driver gear speed (100 rpm) by the driven gear speed (100 rpm). For the gears in figure 15-3B, the speed ratio is:

$$\frac{100 \text{ rpm (driver)}}{50 \text{ rpm (driven)}} = \frac{2}{1} \text{ or } 2:1 \text{ (or } 1:\frac{1}{2} \text{ as indicated)}$$

The ratio of the gears in figure 15-3C is:

$$\frac{100 \text{ rpm (driver)}}{200 \text{ rpm (driven)}} = \frac{1}{2} \text{ or } 1:2$$

In each of these examples, a comparison is made by finding the ratio of speed between the driver gear and the driven gear. As in other types of machines, it is impossible to gain speed and force at the same time.

Ratio of Teeth for Gear Sets and Gear Trains

The number of teeth in a gear determines the speed. The ratio of teeth is found in the same manner as the ratio of speed. The ratio of teeth equals the number of teeth in the driver gear divided by the number of teeth in the driven gear. For the gear pairs in figure 15-3, the ratio of teeth for each set is determined as follows:

(A)	(B)	(C)
Ratio of First Set	Ratio of Second Set	Ratio of Third Set

$$\frac{\text{Teeth in Driver}}{\text{Teeth in Driven}} = \frac{64}{64} \text{ or } 1:1 \qquad \frac{16}{32} \text{ or } 1:2 \qquad \frac{32}{16} \text{ or } 2:1$$

Assume that a certain machine has a gear train consisting of six gears. Three of the gears are drivers having 32, 40, and 48 teeth. These gears mesh with three driven gears having 24, 32, and 40 teeth, respectively. The ratio of teeth for the entire train is the product of the ratios for each set of gears.

$$\frac{\text{Teeth in First Driver}}{\text{Teeth in First Driven}} \text{ or } \frac{32}{24} \times \frac{40}{32} \times \frac{48}{40} \text{ or } 2:1$$

The addition of one, two, or more intermediate gears to any one of the three gear sets does not change the ratio of the driver gear to the driven gear.

Simple and Compound Gear Trains

■ GEAR RATIOS AFFECT FORCE

Gears are an application of the wheel and axle. In addition, gear teeth will mesh only if they have the same size and shape. As a result, if one gear in a gear set has a smaller number of teeth, it must also be smaller in diameter. It follows from the previous statements that the effort force applied through the teeth from the smaller gear to the larger gear can overcome a larger resistance force in the larger gear.

For example, assume that a 16-tooth gear is driving a 32-tooth gear. The smaller gear revolves twice during each revolution of the large gear. The mechanical advantage of force of this gear combination is measured by the ratio of the teeth in the driven gear to the teeth in the driver gear. The MA_f is 32/16 or 2. An effort force of 50 pounds on the 16-tooth gear can overcome a resistance force of 100 pounds on the 32-tooth gear. However, the gain in force is offset by a loss in speed. The speed decreases because the smaller gear must make two revolutions for each revolution of the driven gear.

Compound Gear Trains

A *compound gear train* is formed when a driver gear and a driven gear are affixed to the same stud or shaft and cause another gear or train of gears to turn. A compound gear train provides a greater range of speed change. In addition the gears occupy less space than does a simple gear train. Advantage of this type of gear train is taken in various automotive, electrical, and aviation industries where the speed of motors, engines, or other driving mechanisms must be changed, or where the force must be varied to produce a desired speed or force.

■ ROTATION, SPEED, AND FORCE AFFECTED BY COMPOUND GEAR TRAINS

The gear set illustrated in figure 15-4 is one of the simplest examples of a compound gear train. Gear (A) is a driver turning gear (B). Gear (B) is securely fastened to a shaft so that it causes gear (C) to turn in the same direction and at the same speed as (B). Gear (C), in turn, is the driver for gear (D).

Mechanical Advantage of Speed

In figure 15-4, note that gear (A) turns clockwise, gears (B) and (C) turn

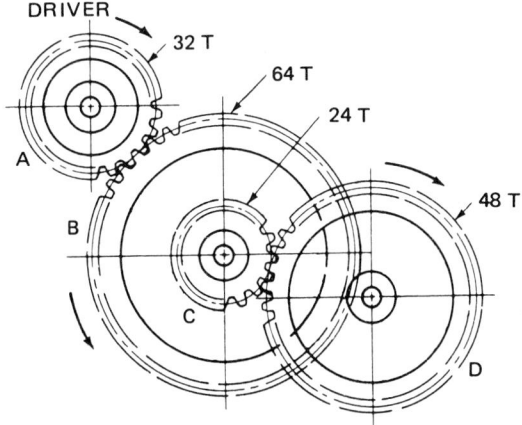

Fig. 15-4 Compound gear train

counterclockwise, and gear (D) turns clockwise. Gear (A) has 32 teeth and turns at 100 rpm. It drives gear (B) (64 teeth) at 50 rpm. Gear (C) also turns at 50 rpm. Since gear (C) has 24 teeth, it drives gear (D) (48 teeth) at 25 rpm. Thus it is possible to reduce the speed by using a combination of gears in different sizes in a compound gear train.

The mechanical advantage of speed for any compound gear train is obtained by multiplying the MA_S of each set of gears.

$$MA_S = \frac{\text{Teeth in First Driver}}{\text{Teeth in First Driven}} \times \frac{\text{Teeth in Second Driver}}{\text{Teeth in Second Driven}}$$

Substituting values from figure 15-4, the mechanical advantage of speed for this gear train is:

$$MA_S = \frac{32}{64} \times \frac{24}{48} = \frac{1}{4}$$

If the initial speed of the first driver gear of a compound gear train is 100 rpm and the MA_S is 1/4, then the final speed is 25 rpm.

■ PRACTICAL APPLICATIONS OF SIMPLE AND COMPOUND GEAR TRAINS

There are many machines for which unusual changes in speeds or feeds are best made by a simple gear train. Some of the older styles of lathes and milling machine dividing heads still use simple gear trains. In those situations where greater compactness is required and the gear ratios are fixed within certain limits, the use of compound gear trains is recommended.

One common example of a compound gear train is the transmission of an automobile, figure 15-5. The letters identify the gears which form the compound gear train.

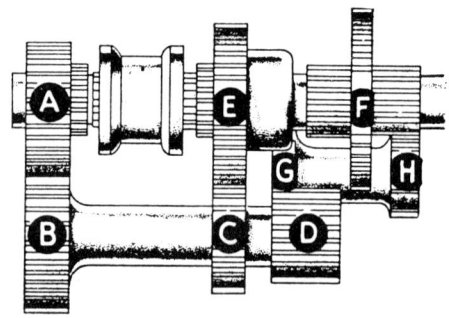

Fig. 15-5

Another use of the compound gear train is in the headstock of certain machines. By meshing gears with different numbers of teeth, a wide range of spindle speeds is possible. Compound gearing is used on quick change gear lathes to provide speed variation between the spindle and the lead and feed screws.

Gear reduction units use compound gear trains to gain tremendous reductions in speed with an equal gain in force that can be transmitted.

SUMMARY

- Gears are used to provide positive rotary motion.
- A simple gear train is a combination of three or more gears driving one another on three or more separate shafts.

- Simple gear trains are used to:
  — change the direction of rotation of shafts,
  — rotate shafts at the same or a different speed, and
  — produce changes in the mechanical advantage of force (MAf) and the mechanical advantage of speed (MAs).
- The mechanical advantage of speed (MAs) of a simple gear train is the ratio of teeth in the driver divided by the number of teeth in the driven gear.
- Compound gear trains can produce the same speed changes as a simple gear train, but they occupy less space and use fewer gears.
- Compound gear trains can be used to change the direction of rotation of shafts as well as the speed of rotation.
- The mechanical advantage of speed (MAs) of a compound gear train is the ratio of the speeds for each set of gears multiplied together.
- Whenever there is a gain in speed in either a simple or compound gear train, this gain is accompanied by a loss in force.

ASSIGNMENT UNIT 15 SIMPLE AND COMPOUND GEAR TRAINS

■ PRACTICAL PROBLEMS WITH SIMPLE AND COMPOUND GEAR TRAINS

Add a word or phrase to complete statements 1-10.

1. Two gears in mesh are referred to as a _____.
2. Gears transmit _____ motion.
3. Gears are used either to _____ or _____ force.
4. Gears may _____ speed.
5. An _____ gear in a simple gear train changes the direction of rotation between the driver gear and the driven gear.
6. The _____ between the driver and driven gears in a simple gear train remains the same regardless of the number of intermediate gears.
7. The ratio of the speed of a simple gear train is found by _____.
8. The relationship between the number of gear teeth in _____ and the number of gear teeth in _____ gear or intermediate gears affects force.
9. One advantage of a gear train is that it is possible to get a _____ speed range in a smaller space.
10. The mechanical advantage of speed of a compound gear train is found by _____ the mechanical advantages (of speed) of each set of gears.
11. Name two industrial applications of (a) a simple gear train, and (b) a compound gear train.

132 Mechanics, Machines, and Wave Motion

12. Prepare a table similar to the one shown.
    a. Mark the directions of rotation of each gear combination A, B, C, and D.
    b. Compute the ratio of speed between the driver gear and the last driven gear in the simple gear trains (A and B).
    c. Find the ratio of speed between the driver gear and the last driven gear in the compound gear trains (C and D).

	Direction of Rotation	Ratio of Speed First Driver to Final Driven Gear
A	64 T, 28 T, 64 T	
B	96 T / 96 T, 48 T / 48 T (compound), then three gears	
C	72 T, 36 T, 48 T, 36 T	
D	32 T / 24 T, 48 T / 64 T, 48 T	

13. Which of the four gear trains in the table for question 12 can deliver the greatest force, if each gear train starts with the same force at the driver gear?

14. Determine the speed in rpm of the driven gear shaft for reach of the four combinations (A, B, C, and D) given in question 12, assuming that the first driver gear revolves at 200, 360, 175, and 480 rpm respectively.

NOTE: The rpm of the last driven gear = rpm of the first driven gear × the ratio of the gear train.

Unit 16 Simple and Compound Machines

Simple machines have three basic functions: (1) they can change the direction of a force, (2) they can change the magnitude of a force, and (3) they can change the speed resulting from a force. Simple machines such as the lever and the inclined plane were studied in previous units as well as their applications to the wheel and axle, the wedge, the cam, the screw, and gears. Each of these machines can perform one or more of the basic functions of simple machines. Other machines such as pulleys and pulley systems and the worm and wheel are also commonly used to perform these functions. This unit covers these latter machines and then applies the principles of simple machines to compound machines.

■ THE PULLEY AND PULLEY SYSTEMS

The pulley is used in construction work, power drives, on industrial and home workshop machines, and in any application where heavy loads must be lifted or moved with a large mechanical advantage. A pulley is another application of the principle of the lever. The word *sheave* identifies a pulley with one or more angular grooves cut from the circumference toward the center. A *single sheave* indicates a pulley with a single groove, *double sheave* means a pulley with a double groove, and so on.

Pulleys may be fixed or movable. A *fixed pulley* stays in one position and does not rise or fall with any movement of the load. A stationary pulley fastened to a crane or some supporting beam is a fixed pulley. The fixed pulley operates on the principle that a resistance force can be moved by the same amount of effort force (neglecting friction). In addition, for the fixed pulley, both the effort and resistance distances are equal. Although fixed pulleys change the direction of a force, they do not increase the mechanical advantage.

Figure 16-1 shows that 50 pounds (22.7 kg) of effort moving downward one foot are required to raise a 50-pound (22.7-kg) resistance force one foot. However, if a movable pulley is used, the same resistance force of 50 pounds (22.7 kg) can be lifted by applying an upward effort force of 25 pounds (11.35 kg). In this case, the effort distance is twice that of the resistance distance. Thus, the mechanical advantage of force is offset by the mechanical advantage of distance.

■ MECHANICAL ADVANTAGE OF THE PULLEY

When several pulleys are used in combination, the assembly is often called a

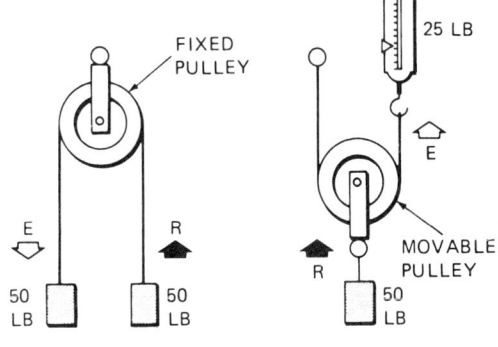

Fig. 16-1

134 Mechanics, Machines, and Wave Motion

block and tackle. This type of pulley combination increases the mechanical advantage of force so that tremendous loads can be lifted with little effort. The mechanical advantage of force is obtained when movable pulleys are used (recall that fixed pulleys change only the direction of the rope or cable from one pulley to the next). The mechanical advantage of force for a block and tackle depends upon the number of ropes supporting the load on the movable pulley.

Figure 16-2 shows a 100-pound (45.4-kg) load being raised with two pulley combinations. Four ropes support the resistance in figure 16-2A. The effort force is 25 pounds (11.35 kg).

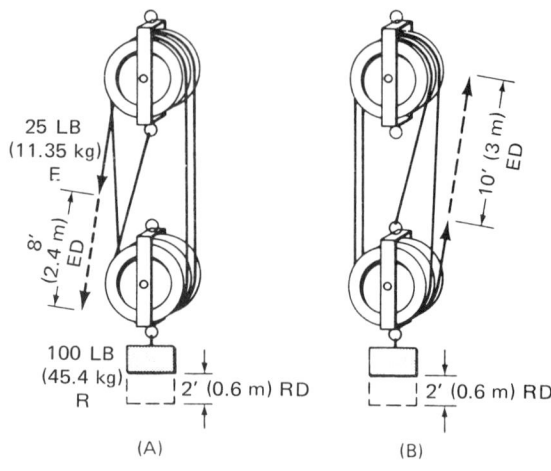

Fig. 16-2

In figure 16-2B, five ropes are supporting the same load. The 100-pound (45.4-kg) resistance in this case, can be moved with an effort force of 20 pounds (9.08 kg). Thus, the load is lifted two feet (0.6 m) when the effort force moves eight feet (2.4 m) in figure 16-2A and ten feet (three meters) in figure 16-2B. In each case, E × ED = R × RD. Any gain in effort is offset by a loss in distance.

■ COMPOUND MACHINES

Simple machines such as the lever and the inclined plane, and the applications of these machines to other machines such as the wedge, the screw, gears, and the block and tackle, can be used alone or in combination with one another. The phrase *complex* or *compound machines* is used to describe two or more simple devices combined in such a way that they work as a single mechanism or device. The result of combining one simple machine with another is that the mechanical advantage of force or speed for the compound machine is the product of the mechanical advantages for each simple machine used.

The Worm and Worm Wheel

The worm and worm wheel is an example of a compound machine which combines the principles of the inclined plane and the wheel and axle. This compound machine can (1) reduce speed, (2) increase force, (3) change the direction of motion, and (4) transmit positive motion.

As a worm turns, its threads engage the teeth on a worm wheel, figure 16-3. For each revolution of the worm, the worm wheel is moved a distance equal to just

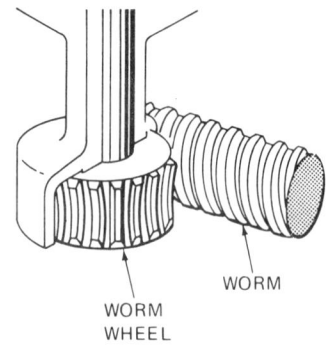

Fig. 16-3

one tooth (for a single pitch screw worm). Figure 16-4 shows a worm wheel with 64 teeth. If this gear is to make one complete revolution, the worm gear must make 64 complete revolutions. If the worm is part of a drive unit turning at 3200 rpm, the shaft to which the worm wheel is attached makes only 50 rpm. When ten pounds (4.54 kg) of effort are applied to turn the worm, the worm wheel can deliver a force of 640 pounds (290.6 kg).

The mechanical advantage of force for a worm and worm wheel combination is the ratio of the number of teeth in the worm wheel to the lead of the worm. For the combination in figure 16-4, the MA_f is 64:1. This fact is important since this type of worm and worm wheel combination is used to multiply effort, decrease speed, and produce motion in parallel planes at an angle to the driver. These conditions are essential for applications such as automotive drives and speed reducers.

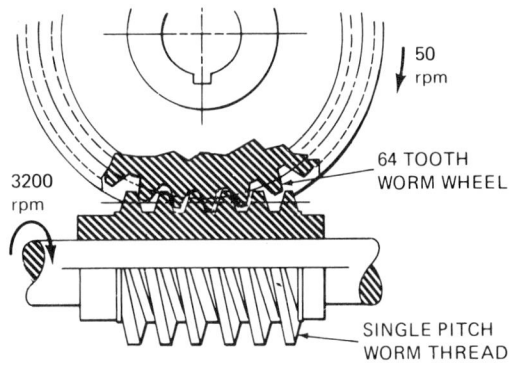

Fig. 16-4

Mechanical Advantage of Compound Machines

The first step in determining the mechanical advantage of any compound machine is to identify the simple machines contained in it. The next step is to determine the mechanical advantage of each simple machine. The product of the mechanical advantages for these simple machines is the mechanical advantage of a compound machine. A common example of a compound machine is shown in figure 16-5. The student should be able to determine that this machine consists of an inclined plane and a pulley system. The mechanical advantage of force (MA_f) for the inclined plane is equal to the slope of the plane divided by the resistance distance, or 20 ft/4 ft (6 m/1.2 m) = MA_f of 5. The MA_f of the pulley system is equal to the number of ropes on the movable pulley, or three. The MA_f of this compound machine is 5 × 3 or 15. As a result, a 300-pound (136.2-kg) effort force can move a resistance of 4500 pounds (2043 kg), neglecting friction.

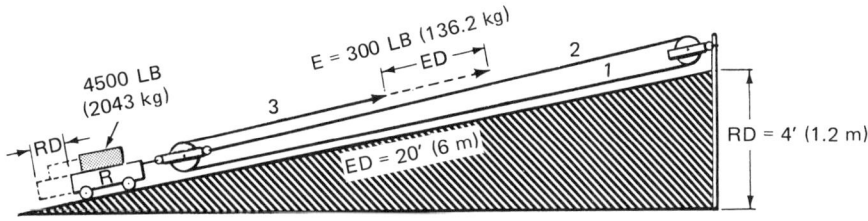

Fig. 16-5

Figure 16-6 shows just one example of a compound machine. The various simple machines can be used in any combination depending upon the amount and kind of work to be done. Both a woodworking vise and a special machine vise are a combination of the inclined plane of the screw thread and the lever (the handle). The automotive clutch and the differential on an automobile are other examples of simple machines combined in different forms to produce the desired motion, changes in forces, and changes in the magnitudes of forces.

Fig. 16-6

================ SUMMARY ================

- A pulley is an application of the lever.

- The mechanical advantage of force of a pulley combination is determined by the number of ropes supporting the load on the movable pulley.

- A gain in the mechanical advantage of force (MA_f) of a pulley combination is offset by a loss in the mechanical advantage of speed (MA_s).

- A compound machine consists of two or more simple machines working in combination with each other.

- Compound machines produce large gains in either force or speed.

- The worm and worm wheel is a compound machine resulting from the combination of the inclined plane and the wheel and axle.

- The worm and worm wheel is used in industry to obtain tremendous reductions in speed and increases in force and to change the angle of rotation.

- The mechanical advantage of force (MA_f) of a compound machine is the product of the mechanical advantages of force (MA_f) of the simple machines combined to form the compound machine.

ASSIGNMENT UNIT 16 SIMPLE AND COMPOUND MACHINES

■ PRACTICAL PROBLEMS WITH SIMPLE AND COMPOUND MACHINES

Pulley and Pulley Systems

1. Match each item in Column I with the correct condition in Column II.

   Column I
   a. Fixed Pulley
   b. Movable pulley
   c. Block and tackle
   d. Mechanical advantage

   Column II
   1. The pulley moves as the load moves.
   2. Depends upon the number of ropes supporting the load on the movable pulley.
   3. Takes more effort to move a smaller resistance.
   4. Several pulleys used in combination with one another.
   5. The pulley remains in one position and does not rise or fall with the load.

2. A 280-kg engine is to be raised one meter and placed on a skid. Two double sheave pulleys are available.
   a. Sketch the pulley arrangement that requires the least effort force to raise the load.
   b. Determine (E) and (ED).

Compound Machines

Select the correct word or phrase to complete statements 1 to 7.

1. A worm and worm wheel is a (compound) (simple) machine.
2. A worm and worm wheel is a combination of these two simple machines: (inclined plane) (pulley) (wheel and axle).
3. A worm and worm wheel is used principally to (increase and decrease) (increase) (decrease) force.
4. A worm and worm wheel is used primarily to (increase) (decrease) speed.
5. A worm and worm wheel produces a MA_f of (one) (more than one).
6. The (product) (sum) of each simple machine in a compound machine combination is the MA_f.
7. A compound machine consists of (a simple machine working alone) (two simple machines working independently of each other) (two or more simple machines combined).

8. From the dimensions and force given in figure 16-7, determine:

   a. the MAf of the gears,
   b. the MAf of the crank,
   c. the MAf of the combined machines, and
   d. the (R) that can be moved

   NOTE: The radius of the small gear is 1 inch.

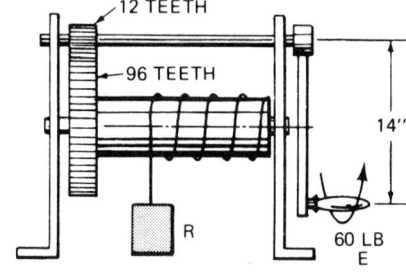

Fig. 16-7

9. A screw jack has a metric thread (M25.4 × 12.7) with a pitch of 12.7 mm, an outside diameter of 25.4 mm, a lever arm with a radius of 35" and an effort force of 75 lb.

   a. Name the simple machines combined in the jack.
   b. Determine the MAf of each simple machine.
   c. Determine the MAf of the combination.
   d. Give the weight of the load that may be lifted, disregarding friction.

10. Study a crane, a derrick, or a pneumatic power shovel on a construction project.

    a. Sketch the pulley combination and also the driving gear mechanism.
    b. Determine the MAf of the fixed and movable pulley combinations.
    c. Find the MAf between the driver and the driven gears on the drive mechanism.
    d. Compute the MAf of the entire compound machine.

Unit 17 Mechanical Power Transmission, Friction, and Lubrication

Power is the term used to indicate the rate of work. Power is equal to the work done divided by the time required to do the work. It follows from this statement that a given amount of work may be done in either less or more time, depending upon the power used.

■ MECHANICAL POWER TRANSMISSION

Power in a usable form is transmitted from the source to the machine or mechanism which produces the desired motion or work. Common sources of power are gas engines, diesel engines, turbines, and electric motors (the most universal of all sources).

Regardless of the source of power, it can be transmitted mechanically from the motor to a battery of machines by means of pulleys and belts. One problem with this method is that the great distances required for power transmission produce large power losses due to belt slippage. Another disadvantage arises because of the excessive weight and size of the entire lineshaft and pulley system.

In many situations, power systems must be designed for locations where there is an obstruction between the driver shaft and the driven shaft. One or more idler pulleys

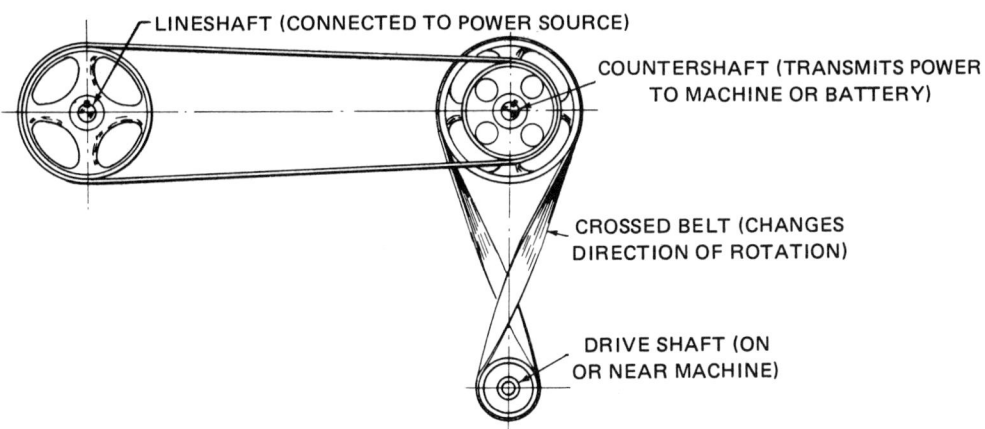

Fig. 17-1 Old "flat belt" system of transmitting power

140 Mechanics, Machines, and Wave Motion

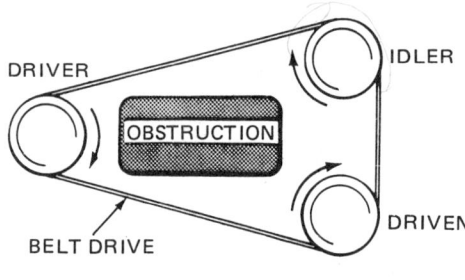

Fig. 17-2

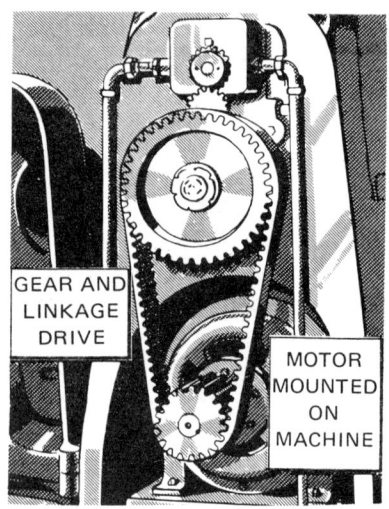

Fig. 17-3

can be inserted in the pulley system to avoid the obstruction, figure 17-2. In theory, the use of idler pulleys does not affect the mechanical advantage of the pulley system.

Modern Mechanical Power Transmission Devices

Direct power transmission in modern industrial equipment means that when a shaft or spindle revolves at a speed controlled within the motor itself or by an electrical device, the power can be delivered directly from the motor shaft. Polishing and grinding machines are examples of the application of direct power transmission.

Modern machine tools combine the mechanical advantage principles of compound machines and have a source of power integral with (included with or a part of) the machine. Such compact machines make it possible to reduce distances and part sizes. As a result there is less power loss than in older machine tools.

The gear is a device commonly used to provide positive power transmission without slippage, figure 17-3.

At higher speeds, a chain drive can be used to produce positive transmission. A chain drive does not require the same high initial tension necessary for either a V-belt or a flat belt.

Power can be transmitted from a device producing a rotary motion to a device requiring a straight line motion by the use of the screw principle. The forward or reverse feed movement of the cutting tool on a metal cutting lathe is controlled by a *feed screw*. This screw transmits rotary power through another part which converts the motion into the straight line motion of the carriage.

Power can be transmitted in the same direction as the motor shaft, or at an angle to the shaft, figure 17-4. A device known as a *coupling* is used to connect the source of power to a given shaft. The *rigid coupling* forms a direct connection or a direct drive. The *flexible coupling* provides a drive at an angle to the power source.

The continuous transmission of power is not always required. Intermittent power use is possible with the use of mechanisms such as clutch arrangements which either engage or disengage the power source.

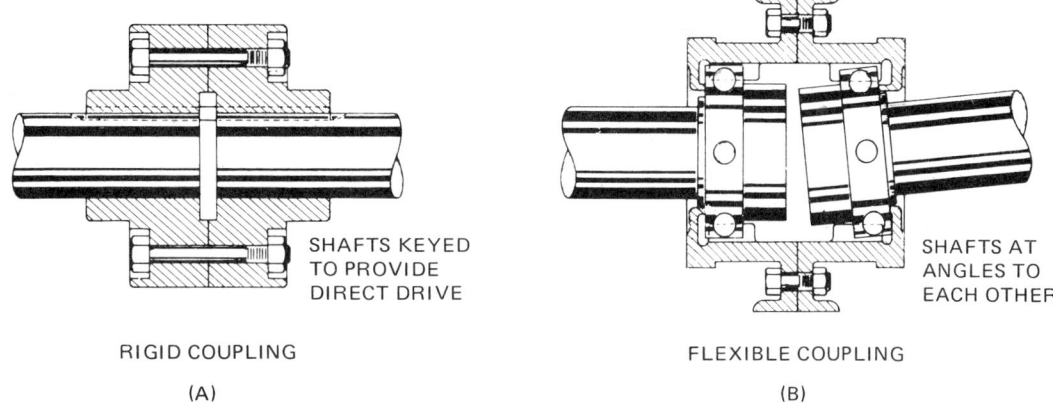

SHAFTS KEYED TO PROVIDE DIRECT DRIVE

SHAFTS AT ANGLES TO EACH OTHER

RIGID COUPLING
(A)

FLEXIBLE COUPLING
(B)

Fig. 17-4

Principles of Mechanical Power Measurement

Reciprocating Machines. The term *reciprocating machine* identifies a combination of parts (machine elements). These parts collectively produce a back and forth, and in and out reciprocating motion. A simple example is a piston which changes the rotary motion of the eccentric cam on a crankshaft into an up and down motion to produce work.

Power was defined as the rate of doing work. The standard measure of power is the horsepower. One horsepower (hp) is equal to 33 000 foot-pounds of work done in one minute (or 550 ft/s, or 746 W). The horsepower required to do a given amount of work in a specified time is given by the following formula. Remember that force × distance = work.

$$\text{hp} = \frac{\text{Force (lb)} \times \text{Distance (ft)}}{33\,000 \text{ ft-lb/min} \times \text{Time (minutes)}}$$

For example, if a reciprocating compound machine exerts a force of 55 pounds over a distance of 100 feet, the horsepower (neglecting friction) is:

$$\text{hp} = \frac{\text{Force (lb)} \times \text{Distance (ft)}}{33\,000 \text{ ft-lb/min} \times \text{Time (min)}} = \frac{55 \text{ lb} \times 100 \text{ ft}}{33\,000 \text{ ft-lb/min} \times 1 \text{ min}} = \frac{5000}{33\,000} = \frac{1}{6}$$

Thus, a motor with a rating of 1/6 hp is required to do this work.

Work, in customary units of power measurement, is given in foot-pounds per second. For practical purposes, a larger unit of power, called the horsepower, is used. In the SI metric system, work is expressed in joules (J). A joule per second is equivalent to one watt (W). Conversions between horsepower and watts are common.

To convert horsepower (hp) to watts (W), multiply the horsepower by 746.

To convert watts (W) to horsepower (hp), multiply the watts by 0.001 34.

Example: Calculate (a) the power (W) developed and (b) the horsepower of an engine having the following specifications. (Friction and other power losses are neglected.)

- Average force exerted on an engine piston during the power stroke = 800 newton (N)
- Length of power stroke = 127 mm
- Power strokes per second = 12
- Number of cylinders = 6

① Calculate the work for one second.

Work (W) = Force (N) × Distance (in meters)

= 800 N × 0.127 m

= 101.5 N·m/stroke

Since 1 newton (N) = 1 joule (J), the (W) of 101.5 N/stroke = 101.5 J/stroke.

② Calculate Power for one second.

$$\text{Power} = \frac{\text{Work} \times \text{Number of Cylinders} \times \text{Strokes per Second}}{\text{Time (Seconds)}}$$

$$= \frac{101.5 \times 4 \times 12}{1} = 4872 \text{ J/sec} = 4872 \text{ W}$$

③ Calculate the horsepower.

HP = Watts × 0.001 34

= 4872 × 0.001 34

= 6.5 hp

Rotating Machines (Prony Brake). *Rotating machines* consist of machine parts or elements which produce a turning or rotary motion. The flywheel of an engine, the armature shaft of an electric motor, and the spindle of a power tool are classified as rotating machinery. All of the machine elements just listed are actuated or controlled by other machine parts.

The mechanical power of a rotating machine can be measured with an instrument known as a *dynamometer*. This device absorbs the energy output of the machine, and converts it to heat or electrical energy which can be measured in watts. One type of dynamometer is the Prony brake, which measures the brake horsepower delivered to the flywheel shaft, or spindle, of a machine. A simple type of *Prony brake* consists of a band or belt that passes around a revolving pulley on the power source. The ends of the belt are secured to spring balances.

The load on the machine being tested is regulated by two adjusting nuts. The forces exerted on the belt are indicated by the readings on the spring balances. The frictional force of the belt on the Prony brake is equal to the difference between the two spring balance readings.

The amount of work done per minute (power) by the machine during each revolution to overcome brake friction is equal to the difference in readings multiplied by the circumference of the pulley. For one minute this amount of work is again multiplied by the number of times the machine revolves in one minute. Expressed as a formula, the work (W) done by the machine is equal to:

$$W = (F_1 - F_2) \times (\text{Circumference of the pulley, } \pi D) \times (N)$$

where N = revolutions per minute

F_1, F_2 = readings on the spring balances

D = diameter of the pulley

If the F_1 and F_2 readings are in pounds, and the circumference is given in feet, W is given as foot-pounds. The brake horsepower can be determined by dividing the amount of work done by the machine (in foot-pounds) by the 33 000 foot-pounds per minute.

$$hp = \frac{(F_1 - F_2) \times (\pi D) \times (N)}{33\,000 \text{ ft-lb/min}}$$

Figure 17-5 shows a difference of 9 pounds between F_1 and F_2 acting on a 7-inch diameter pulley turning at 1000 rpm. Using the Prony brake formula and substituting values, the brake horsepower of the machine being tested can be obtained.

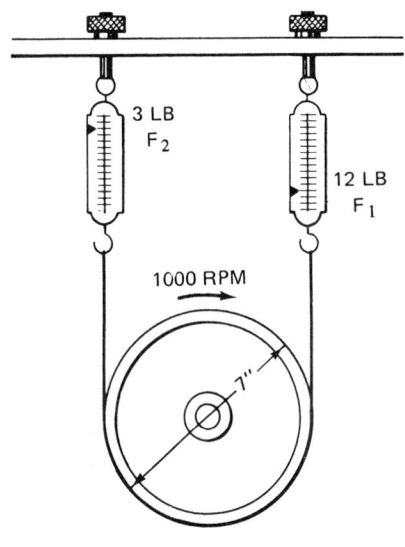

Fig. 17-5

$$hp = \frac{(12 \text{ lb} - 3 \text{ lb}) \times \left(\frac{22/7 \times 7 \text{ in}}{12 \text{ in}}\right) \times (1000 \text{ rpm})}{33\,000 \text{ ft-lb/min}} = \frac{1}{2}$$

The Prony brake provides a fast and simple method of measuring the mechanical power of rotating machines.

■ FRICTION: CAUSES AND TYPES

In many of the problems thus far, the student has been asked to disregard or neglect friction. Actually, friction is present to some degree whenever two parts are in contact and move on each other. The term *friction* refers to the resistance of two or more parts to movement.

Valuable or Harmful Friction

Friction is harmful or valuable depending upon where it occurs. Friction is necessary for fastening devices such as screws and rivets which depend upon friction to hold the fastener and the parts together. Belt drives, brakes, and tires are additional applications where friction is necessary.

The friction of moving parts in a machine is harmful because it reduces the mechanical advantage of the device. The increased effort force required to overcome friction between the turning parts causes a loss in power and a resulting increase in heat.

The heat produced by friction is lost energy because no work takes place. Also, greater power is required to overcome the increased friction. Heat is destructive in that it causes expansion. Expansion may cause a bearing or sliding surface to fit tighter. If a great enough pressure builds up because of expansion, the bearings may *seize* or freeze. In addition, as heat is developed, bearings made from low temperature materials may melt.

Causes of Friction

Friction in solids results from any one or combination of three general causes: (1) surface finish, (2) the cohesion or adhesion of molecules, and (3) weight or pressure.

Surface Finish. A microscopic examination of two surfaces in contact with each other shows irregularities in both surfaces, figure 17-6. As the two surfaces move, the high and low points of each tend to interlock and prevent the surfaces from sliding freely over each other. As a result, a great force is needed to start or keep the parts moving. The surface irregularities cause a grinding or shearing action as the surfaces move over each other. This wear produces heat and subsequent power losses.

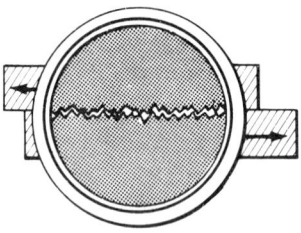

Fig. 17-6

Intermolecular Attraction. Friction is believed to be caused by *cohesion*, or the attraction of like molecules of the same material. Cohesion is greater as the surfaces of two like materials come closer in contact. Thus, for two like materials with highly polished surfaces, the following is true of the friction: (1) the friction caused by surface roughness is greatly reduced, and (2) the friction caused by cohesion is increased.

Friction is also caused by *adhesion*, or the attraction of unlike molecules in two different surfaces. The tendency of some soft metals to cling to rotating parts is due to adhesion. Similarly, the attraction between liquids and the walls of piping is the result of adhesion.

Types of Friction

There are three types of friction which must be overcome in moving parts: (1) starting, (2) sliding, and (3) rolling. *Starting friction* is the friction between two solids that tend to resist movement. When two parts are at a state of rest, the surface irregularities

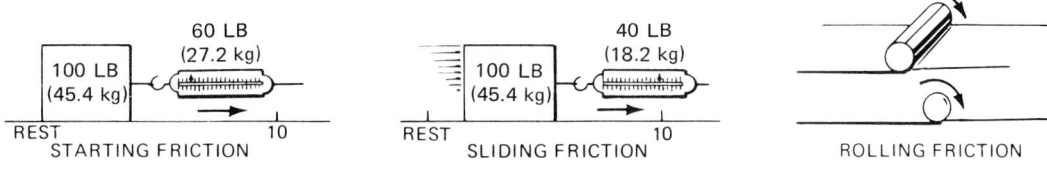

Fig. 17-7

of both parts tend to interlock and form a wedging action. To produce motion in these parts, the wedge-shaped peaks and valleys of the stationary surfaces must be made to slide out and over each other. The rougher the two surfaces, the greater is the starting friction resulting from their movement.

Since there is usually no fixed pattern between the peaks and valleys of two mating parts, the irregularities do not interlock once the parts are in motion but slide over each other. The friction of the two moving parts is known as *sliding friction*. As shown in figure 17-7, starting friction is always greater than sliding friction.

Rolling friction occurs when roller devices are subjected to tremendous stresses which cause the parts to change shape or *deform*. Under these conditions, the material in front of a roller tends to pile up and forces the object to roll slightly uphill. This changing of shape, known as *deformation*, causes a movement of molecules. As a result, heat is produced from the added energy required to keep the parts turning and overcome friction.

■ COEFFICIENT OF SLIDING FRICTION

The force required to overcome friction depends upon the forces pressing the two surfaces together. These two forces are directly proportional. If a 100-pound force is required to slide a load, a 200-pound force is needed if the load is doubled. Similarly, a 300-pound force is needed if the load is tripled, and so on, figure 17-8.

The ratio of the friction force resisting motion and the perpendicular force pressing the two surfaces together is known as the *coefficient of friction*. When the object is on a horizontal surface and the force pressing the two surfaces together is perpendicular to the horizontal surface, then the coefficient of friction is expressed by the formula:

$$\text{Coefficient of Friction} = \frac{\text{Force (causes object to slide)}}{\text{Weight (of object)}}$$

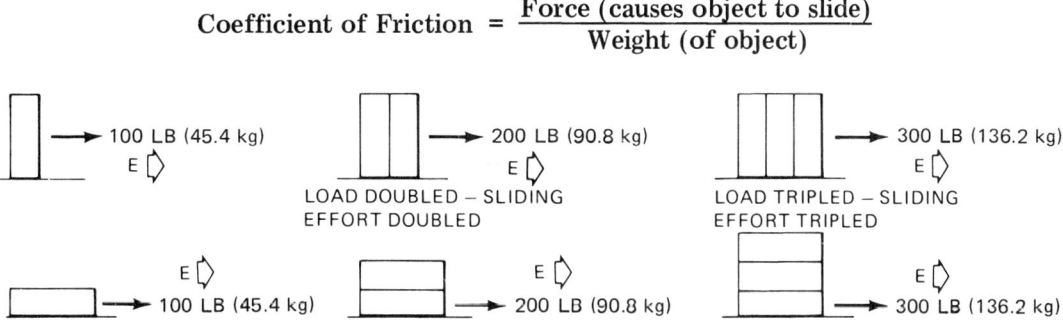

Fig. 17-8 Effort forces for same materials are proportional to load

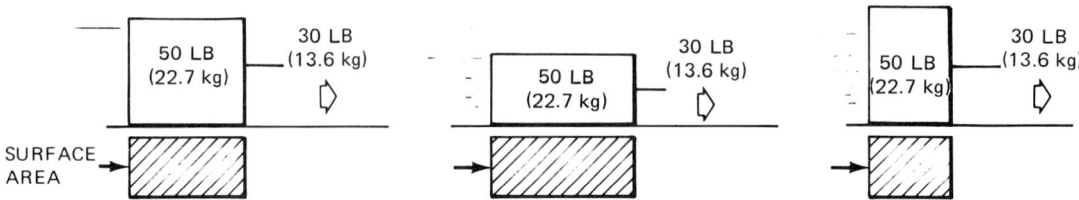

Fig. 17-9 Sliding friction of same material is unaffected by surface area

For example, if 50 pounds (22.7 kg) are required to move a 100-pound (45.4-kg) block along a horizontal surface, the coefficient of friction is equal to force/weight: $\frac{50}{100} = 0.5$ (or $\frac{22.7}{45.4} = 0.5$).

Since values of coefficients of friction are used so often, tables are available which give approximate values for pressures ranging between 14 and 20 pounds (or 6.4 to 9 kg) per square inch for various materials. Coefficients of friction must be increased for standing or starting friction and when higher pressures are used. A coefficient of friction table shows clearly those materials that may be used together when friction is needed. In addition, combinations of materials are shown which can be used when parts must slide freely over one another. When coefficient of friction tables are available and the weight of a sliding part is known, it is possible to determine the force pressing two surfaces together using the formula for the coefficient of friction.

Friction is considered independently of surface area. The friction between two materials remains the same regardless of the surface area (if the area is constant).

Figure 17-9 illustrates three objects of equal weight and material. Each object requires the same force to cause it to slide on the horizontal surface.

■ PRINCIPLES INVOLVED IN REDUCING FRICTION

The friction caused by the wedging action of surface irregularities can be overcome partly by the precision machining of the surfaces. However, even these smooth surfaces may require the use of a substance between them to reduce the friction still more. This substance is usually a lubricant which provides a fine, thin oil film. The film keeps the surfaces apart and prevents the cohesive forces of the surfaces from coming in close contact and producing heat.

Another way to reduce friction is to use different materials for the bearing surfaces and rotating parts. This explains why bronze bearings, soft alloys, and copper and tin oilite bearings are used with both soft and hardened steel shafts. The *oilite* bearing is porous. Thus, when the bearing is dipped in oil, capillary action carries the oil through the spaces of the bearing. This type of bearing carries its own lubricant to the points where the pressures are the greatest.

Overcoming Friction with Bearings

Roller and ball bearings are used to reduce friction by substituting a rolling action for a sliding action. The roller or ball bearing moves over the surface irregularities and

Mechanical Power Transmission, Friction, and Lubrication   147

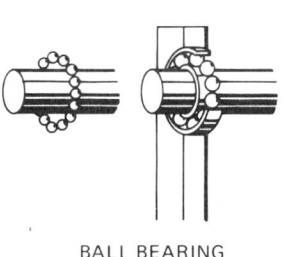 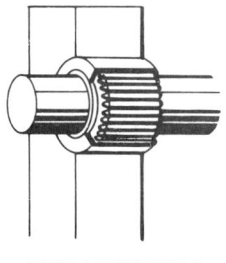 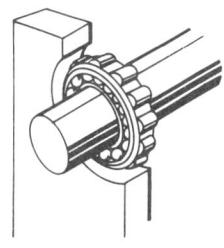

BALL BEARING   NEEDLE BEARING   ROLLER BEARING

Fig. 17-10

does not form a wedge with the irregularities as in the case of flat surfaces. There are many different uses for roller and ball bearings. If greater loads are to be supported, the roller bearing is preferred because the area of contact is greater than with the single point of contact obtained with a ball bearing, figure 17-10.

A needle bearing is also commonly used to overcome friction. The needle bearing is similar to the roller bearing except that the rollers are smaller and there are a greater number of them as compared to a roller bearing. A needle bearing has an advantage over the roller bearing in that for the same load-carrying capacity, the needle bearing is lighter and better adapted to machines and mechanisms where weight is important.

Fig. 17-11

Ball and roller bearings are supported and spaced uniformly by a frame. The friction caused by the movement of the bearings in the frame can be overcome by lubrication. However, if the load on a bearing is increased to a point where the race in which the bearing rolls tends to become flat at the point of contact, the material piles up in front of the roller, figure 17-11. This effect creates molecular friction which in turn produces heat. This condition cannot be corrected by lubrication. Either the shape or size of the bearings or the construction of the mechanism must be changed to overcome this condition.

■ PRINCIPLES OF LUBRICATION

Moving parts are lubricated to reduce friction, wear, and heat. The most commonly used lubricants are oils, greases, and graphite compounds. Each lubricant serves a different purpose. Essentially, a lubricant must meet three conditions:

(1) It must have sufficient body to be able to stand up under heavy loads or the operating conditions of the machine on which it is used. For example, since most mechanisms heat up during operation, an oil used as a lubricant must be able to withstand the heat of the unit and still provide the lubrication needed for continuous operation.

148 Mechanics, Machines, and Wave Motion

(2) The lubricant must flow freely so that a thin film is maintained between the moving parts. A lubricant that flows too slowly and does not easily reach all parts causes undue heat and wear.

(3) The composition of the lubricant should prevent chemical corrosion. Usually, a thin coating of oil is sufficient to prevent the formation of rust.

Methods of Lubrication

The conditions under which two moving surfaces are to work determine the type of lubricant to be used and the system selected for distributing the lubricant. On slow moving parts with a minimum of pressure, an oil groove is usually sufficient to distribute the required quantity of lubricant to the surfaces moving on each other.

A second common method of lubrication is the splash or bath system in which parts moving in a reservoir of lubricant pick up sufficient oil which is then distributed to all moving parts during each cycle. This system is used in the crankcases of lawn-mower engines to lubricate the crankshaft, connecting rod, and parts of the piston.

A lubrication system commonly used in industrial plants is the pressure system. In this system, a pump on a machine carries the lubricant to all of the bearing surfaces at a constant rate and quantity.

There are numerous other systems of lubrication and a considerable number of lubricants available for any given set of operating conditions. Modern industry pays greater attention to the use of the proper lubricants than at any other previous time because of the increased speeds, pressures, and operating demands placed on equipment and devices.

SUMMARY

- Power is the rate of doing work. Power = Work ÷ Time:

$$\text{Power} = \frac{\text{Work (in foot-pounds)}}{33\,000 \text{ ft lb/min} \times \text{Time (in min)}},$$

 expressed in units of horsepower. Horsepower values may be converted to equivalent watt values by multiplying by 746.

- Positive power transmission is provided by using gears, linkages and other mechanical devices, or by direct drive.

- Power losses are reduced when the source of power is brought as close as possible to the point of drive.

- Generally, friction is caused by surface finish, the cohesion and adhesion of the molecules of the moving parts, and the weight or pressure on the surfaces.

- Three common types of friction are: starting, sliding and rolling.

- The coefficient of friction is the ratio between the force required to move an object and the weight of the object.

- Friction due to surface irregularities can be reduced by precision machining, changing the materials in the parts that move, using ball and roller bearings, or by the use of lubrication.
- Lubricants are used to reduce heat, wear, and friction.
- Lubricants must have: sufficient viscosity to prevent moving surfaces from coming into contact, the ability to flow easily and to stand up under continued service without breaking down or becoming gummy, and freedom from rust-producing impurities.

ASSIGNMENT UNIT 17 MECHANICAL POWER TRANSMISSION, FRICTION, AND LUBRICATION

■ PRACTICAL PROBLEMS IN POWER MEASUREMENT, FRICTION, AND LUBRICATION

Mechanical Power Transmission and Measurement

For statements 1 to 8, determine which are true (T) and which are false (F).

1. Power and work are the same.
2. Power refers to the number of foot-pounds of work done.
3. Power can be transmitted by direct drive using rigid or flexible couplings.
4. Gears and linkage systems provide less positive power transmission than flat belts.
5. One horsepower = 33 000 foot-pounds of work per minute.
6. One watt = 746 hp.
7. A clutch is preferred to a coupling in transmitting power continuously without interruption.
8. The Prony brake is a simple measuring instrument for determining the brake horsepower of rotating machines.
9. The results of Prony brake tests of three engines are given in the table. Compute the brake horsepower of each engine to the nearest 1/4 hp.

	Force (in lb)		Pulley	
	F_1	F_2	Diameter (inches)	Rpm
Engine A	5	9 1/2	3 1/2	6000
Engine B	75	152	12	680
Engine C	10	259	14	903

Friction

Select the correct word or phrase to complete statements 1 to 8.

1. Of the two types of friction, (sliding) (starting) friction is always greater.
2. As the weight of a moving part increases, the friction (remains the same) (decreases) (increases).
3. In general, two rough surfaces require a (larger) (smaller) force to move one over the other than two smooth surfaces.
4. If the weight of a part remains the same, the surface area (affects) (does not affect) friction.
5. The coefficient of friction is the ratio of the (effort force ÷ weight of the object) (weight of the object ÷ effort force).
6. Roller and ball bearings (decrease) (increase) friction.
7. Roller and ball bearings work on the principle of (sliding) (starting) (rolling) friction.
8. The (needle) (roller) (ball) bearing is best suited for heavy loads where light weight and the strength of the bearing and retainer are important.
9. List three common causes of friction.
10. List three common types of friction.
11. Study a car manufacturer's recommendations for breaking in a new car. Give five reasons for supporting these recommendations.
12. Name five parts or mechanisms in an automobile that depend upon friction for safe operation.
13. List five parts or mechanisms in an automobile where friction is undesirable.
14. Examine the surface finish of four different materials with a microscope. Make simple sketches of the surfaces and exaggerate the surface finish. Number the parts from 1 to 4 and range them in order from the coarsest to the smoothest, respectively.
    a. Which of the four surfaces provides the best friction?
    b. Which surface, in terms of surface finish, will wear the least and have the smallest starting friction?
15. Sketch the enlarged surfaces of two roughly machined parts that move on each other. Explain why starting friction is greater than sliding friction.
16. Wooden and metal rollers are used to move heavy equipment easily. Explain briefly why rollers are used.

17. The data from friction tests using different materials are given in the accompanying table. Determine the coefficient of sliding friction for combinations A, B, C, D, and E.

Materials	A*	B*	C*	D*	E*
	Leather on Cast Iron	Leather on Hardwood	Bronze on Bronze	Cast Iron on Hardwood	Cast Iron on Cast Iron
Force	140 lb	111 lb	215 lb	140 kg	20 kg
Weight of part or load	250 lb	335 lb	1075 lb	285 kg	133 kg
Coeff. friction (sliding)					

*A, B, C, and D are dry; E has slight lubrication.

18. Arrange the materials in the table in order from 1 to 5 to indicate the lowest to the highest coefficients of friction, respectively. Use the letters to designate the materials.
19. What combination of materials from the table should be used in parts which depend upon friction for their operation?
20. What two materials from the table are best suited for parts that must slide continuously?

Lubrication

For statements 1-8, determine which are true (T) and which are false (F).
1. Lubricants are used to reduce starting and sliding friction.
2. In industry, careful attention is given to lubricants that have rust-prevention qualities.
3. The materials used to make moving parts have no influence on the type of lubricant needed.
4. The splash system of lubrication depends on the movement of parts in a reservoir of oil.
5. A slow moving lubricant always provides a sufficient flow to lubricate all parts.
6. The pressure or forced feed system of lubrication is the least expensive to install and is used on parts that move slowly under light loads.

7. Lubricants reduce cohesion and adhesion by providing a film of liquid between moving parts.
8. The lubrication of moving parts has no effect on the amount of heat produced.
9. Select the letters in the following list of those items that require lubrication.
    a. Flat leather belt
    b. Flat bearing surface
    c. Machine spindle
    d. V-belt
    e. Shaft bearing
    f. Brake band
10. Identify two different types of roller or ball bearings and two other regular bearings. Name two parts or machines and two conditions where the four bearings can best be used.
11. Explain why it is easier to start a heavy drive after a momentary stoppage than it is when the shaft remains stationary for a long time.
12. Why are lighter oils used in aircraft, automotive, and diesel engines in cold working temperatures?
13. Give three reasons why the used oil in the crankcase of an engine should be drained off periodically and replaced.
14. Give a brief explanation of why the use of a lubricant on two moving parts reduces wear, heat, and friction, and lengthens the life of a machine.

Unit 18 Mechanics of Fluids at Rest and in Motion

A liquid has no definite form but takes the shape of its container. Similarly, a gas has neither form nor a definite volume. It, too, takes the shape of the vessel or container. In both cases, the word fluid can be applied to either the gas or the liquid because both substances can flow. The energy of the liquid or gas in a sealed container can be harnessed to do useful work.

This unit gives the basic principles and properties of fluids. Common applications of fluids to solve everyday needs are also provided. The unit begins with a review of the simple terms of pressure, density, buoyancy, and specific gravity as each of these properties apply to fluids.

■ FLUID PRESSURE AND FORCE

Pressure refers to the force per unit area. The result is usually expressed as a pressure of so many pounds per square foot (lb/ft^2), or as grams per square centimeter (g/cm^2), or as any other combination of weight and area measure. If five tons of concrete are poured over an area ten feet square, the pressure per square foot = 10 000 lb ÷ 10 ft × 10 ft, or 100 lb/ft^2.

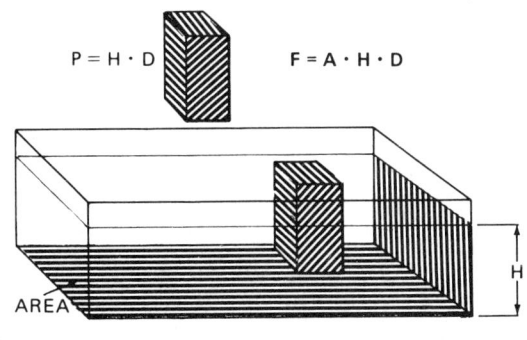

Fig. 18-1

There are times when the total force must be determined. At other times, the pressure for a unit area must be known. The force on the bottom of a tank is equal to the volume of the fluid × the density of the fluid.

For a rectangular or square tank, figure 18-1, the volume is the area of the base multiplied by the height of the fluid.

$$\text{Force (F)} = \text{Area (A)} \times \text{Height (H)} \times \text{Density (D)}$$

In addition,

$$\text{Pressure (P)} = \text{Force (F)} \div \text{Area (A)}$$

If the value of (P) is substituted in the force equation, the areas (A) cancel out and (P) is equal to (HD).

$$F = AHD \text{ and } P = \frac{F}{A}$$

Therefore,

$$P = \frac{AHD}{A} = HD$$

153

difference between pressure and force is that pressure depends upon the area ...act and force depends upon the volume area and density.

For example, compare the difference between the pressure and the force at the ...tom of a square tank five feet on a side. The tank is filled with gasoline to a height of 10 feet (gasoline weighs 42 lb/ft^3). The pressure at the bottom of the tank = HD = 10 ft × 42 lb/ft^3 = 420 lb/ft^2 (or 190.7 kg/ft^2).

For the same tank, the force (F) is equal to (AHD): F = 5 ft × 5 ft × 10 ft × 42 lb/ft^3 = 10 500 pounds (4767 kg). Fluid pressure is expressed as a weight per area such as 420 lb/ft^2; fluid force is expressed by the total number of pounds, such as 10 500 lb (4767 kg).

Experimentation has shown that the pressure at a given depth in a fluid is the same in all directions. If the pressure is greater in any one direction, then the fluid will flow toward that area. This fact was first stated by Pascal and later became known as *Pascal's Law*. The law applies to all enclosed fluids at rest, both gaseous and liquid. Under these conditions the law states that any pressure applied to an enclosed fluid is transmitted equally in every direction without loss. In addition, the fluid acts with equal force on all surfaces.

The action of this law is demonstrated by a special glass bottle as shown in figure 18-2. As pressure is applied to the liquid by the piston, the liquid is sprayed at the same pressure and in the same amounts from each of the small holes, regardless of the positions of the holes.

Assume that this same action is duplicated with a metal container having a number of low-pressure gas gages attached. When air is pumped into the container, each gage registers the same pressure. For both the glass bottle and the metal container, the fluids are at rest and enclosed in a container.

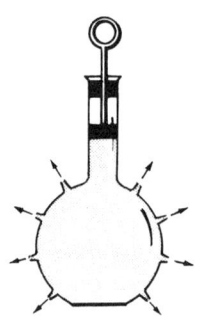

Fig. 18-2

■ PRESSURES ON FLUIDS ARE UNAFFECTED BY SHAPE AND SIZE OF CONTAINER

The size and shape of a vessel have no effect on the gravity pressure of the fluid. *Gravity pressure* refers to the pressure exerted by the fluid itself in the absence of any additional external pressure. The fluid exerts pressure against any surface that it touches.

The fact that the size and shape of a container do not affect the pressure of a fluid can be demonstrated by three or four glass containers of different sizes, figure 18-3. Each container has the same neck size. If each container is filled to the same height with the same liquid, the neck of each is covered with a rubber diaphragm, and then each container is mounted in a test stand, the indicator readings for the pressures at the bottom of each vessel are identical. Note that the volume of liquid in each

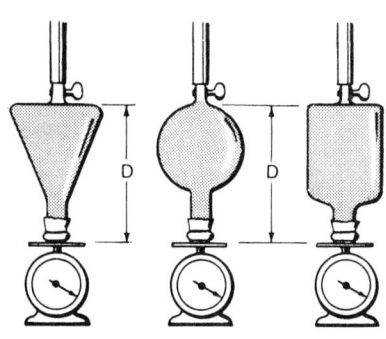

Fig. 18-3

Mechanics of Fluids at Rest and in Motion   155

container and the container shapes are different. However, the height of the fluid in the container and the density of the liquid are constant. When these values are substituted in the formula P = HD, the value of P remains the same for each container.

This same principle governing the action of fluids is used in engineering in the building of dams. Since pressure increases with depth, reservoir and power dams are built to withstand greater pressures at the bottom of the dam than at the top surface.

In large communities, water supply systems depend upon these same principles to insure that water is delivered at the same pressure to all homes.

■ APPLICATION OF FLUID PRESSURE TO MACHINES: HYDRAULIC PRESS

Practical applications of Pascal's Law are found wherever the energy of a fluid can be harnessed to produce useful work. Hydraulic jacks, air drills, automobile tires, and hydraulic presses are but a few of the applications. One of the widest uses of the principles of fluid pressure is in the hydraulic press. The operation of this device depends upon the fact that pressure in a liquid at rest is transmitted unchanged (except for changes in level) in all directions. Hydraulic presses are capable of exerting enormous forces.

A simple hydraulic press is shown graphically in figure 18-4. The press consists of a large piston and a small piston which each have a watertight fit in the press housing. The two pistons are connected by a small pipe. Two valves are provided so that additional liquid can be drawn into the small valve chamber on the upstroke while the valve in the large piston chamber is closed. As the small piston begins its downward stroke, the supply valve closes and the valve to the large piston chamber opens. In this manner, the pressure exerted by the small piston is transmitted to the large piston.

Mechanical Advantage of the Hydraulic Press

If both pistons of the hydraulic press are the same size, the force transmitted by one piston is equal to the force of the other piston. In figure 18-4, one piston is 100 times larger than the small piston. Therefore, a ten-pound (4.54-kg) effort on the small piston produces a force of 1000 pounds (454 kg) on the large piston. The mechanical advantage of this combination (neglecting friction) is equal to R ÷ E = 1000 ÷ 10 = 100 (or 454 ÷ 4.54 = 100). In other words, the forces on both pistons are proportional to the areas of the pistons.

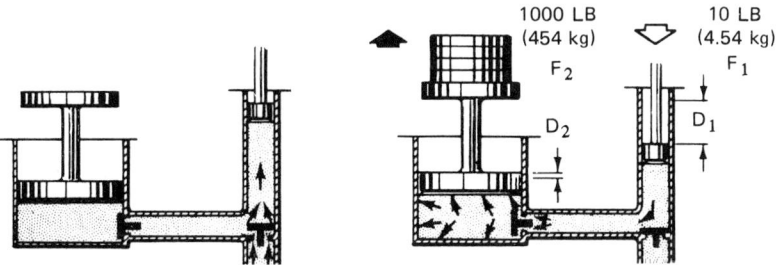

Fig. 18-4

The increase in force from the small to the large piston must be compensated for by a loss in distance. One of the fundamental laws for an ideal machine is that one force and the distance through which it acts is equal to another force and the distance through which this second force acts on the same machine.

$$\text{Force}_1 \times \text{Distance}_1 = \text{Force}_2 \times \text{Distance}_2$$

Thus, to compensate for the increased force on the large piston of the hydraulic press, the distance through which the small piston moves is increased. If the mechanical advantage of force (MA_f) = 100, the small piston travels 100 times farther than the large piston.

The hydraulic press is used in the automotive, aircraft, and structural industries to produce the enormous forces necessary to manufacture heavy forgings, car bodies, fenders, and structural shapes.

■ SPECIFIC GRAVITY MEASUREMENTS OF FLUIDS

In one of the early units on the properties of liquids, Archimedes' principle of buoyancy was discussed. A fluid exerts an upward buoyant force upon a body that is wholly or partly submerged in it. Thus, the amount of the force is equal to the weight of the fluid displaced. The buoyancy is affected by the specific gravity of the fluid. Recall that the specific gravity of a body is the ratio of the weight of the body to the weight of a standard substance having the same volume as the body.

For liquids, the standard substance is water at 39.8°F. At this temperature, water is at its maximum density. The specific gravity of a sinking (heavy) body, of a floating body, or of a liquid is found by the following basic formulas:

$$\text{Specific Gravity} = \frac{\text{Weight of Body}}{\text{Weight of Equal Volume of Water}}$$

$$= \frac{\text{Weight of Body}}{\text{Loss of Weight in Water}}$$

Specific gravity is expressed as a numerical value. Specific gravity values that are larger than one (1.00) indicate that the body is heavier than water; those values under one indicate that the body is lighter than water. A later unit of this text covers the standard substance on which the specific gravity system of gases is based.

Specific gravity can be computed or it can be determined with an instrument known as a *hydrometer*. The hydrometer works on Archimedes' principle. Hydrometers usually consist of a weighted glass bulb. When the bulb is placed in a liquid, it sinks to the depth at which it is displacing exactly its own weight of the liquid. The specific gravity of the liquid determines how far the bulb will sink. The lower (smaller) the specific gravity of the liquid, the lower the bulb sinks in the liquid. A calibrated scale mounted on the bulb provides a simple method of reading the specific gravity of the liquid.

■ FLUID FLOW, PRESSURE, AND SPEED: BERNOULLI'S PRINCIPLE

Pascal's law deals with fluids at rest in confined vessels where an external pressure is transmitted equally in all directions. As the fluid starts to move, the pressure is no

Mechanics of Fluids at Rest and in Motion 157

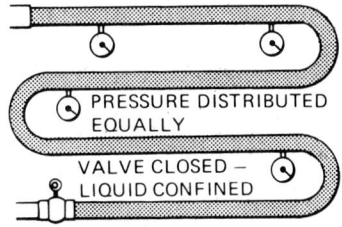
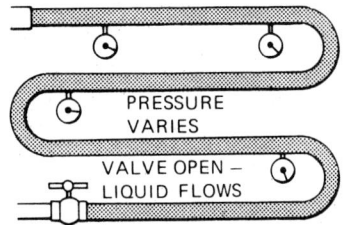

Fig. 18-5

longer the same in all parts of the vessel because a force is necessary to overcome friction. This fact is demonstrated by the bent copper tubing shown in figure 18-5. When the valve is closed and water fills the tubes, each gage indicates the same pressure because the water is confined and the pressure is exerted equally in all directions. However, when the valve is opened, the gages show a difference in pressure due to the friction within the tube. Thus, when liquids are to be piped for great distances, pumping stations must be installed at intervals so that a constant pressure can be maintained throughout the entire line.

Bernouilli's principle states that whenever the velocity of a fluid is increased at any point, its pressure is decreased. For example, if a stream of air is blown across the top of a glass tube dipped in a liquid, the liquid will rise in the tube. This action is due to the fact that the air flowing across the tube decreases the atmospheric pressure at the top of the tube. If the velocity of the air is great enough, the liquid is drawn completely up the tube and is blown with the passing air in a fine spray. Common applications of Bernoulli's principle include spray guns, atomizers, and carburetors. These devices act like suction pumps with no moving parts.

Another example of Bernoulli's principle is demonstrated when a pipe is reduced in cross-sectional area (*constricted*). In this case, the speed of the liquid through the pipe increases at the reduced point so that the same amount of liquid may pass through the constriction as through any other portion of the pipe. This increase in speed causes a decrease in pressure at this point, figure 18-6.

The venturi gage is based on this latter application of Bernoulli's principle. The venturi gage is a tubelike device with a constricted portion. Fluid passing through the venturi increases in speed as it passes through the narrowed portion. At the same time, the pressure is reduced at the narrowed portion. This pressure reduction can be measured, and the rate of flow can be determined or controlled.

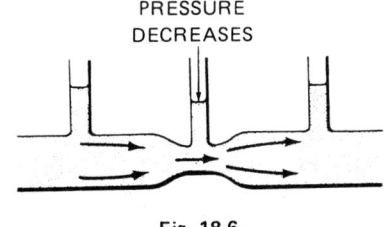

Fig. 18-6

SUMMARY

- The term *fluids* refers to both gases and liquids. The molecules of a fluid are free to move or to be pushed in all directions.

- Pressure is a force per unit area. Pressure is expressed in weight and area measure: P = HD.
- Force is a measure of area, height of fluid, and density: F = AHD. Force is expressed in weight measure.
- Pascal's law for confined fluids at rest states that any pressure applied to the fluid is transmitted in all directions without loss.
- Gravity pressure is the pressure exerted internally by the weight of the fluid itself and is not affected by the size and shape of the container.
- Machines such as the hydraulic press, hydraulic brakes, and pumps, are practical applications of the scientific principles of force, distance, and pressure combined to do productive work.
- At times, the specific gravity measurement of a fluid must be determined before calculations for pressure or force can be made.
- The specific gravity of sinking bodies, floating bodies, and liquids equals the weight of the body divided by the weight of an equal volume of water.
- The pressure of fluids in motion drops due to friction.
- Bernoulli's principle states that whenever the speed of a fluid is increased, the pressure decreases.

ASSIGNMENT UNIT 18 MECHANICS OF FLUIDS AT REST AND IN MOTION

■ PRACTICAL PROBLEMS WITH FLUIDS AND PRESSURES

Fluid Pressure and Force

Select the correct word or phrase to complete statements 1 to 6.

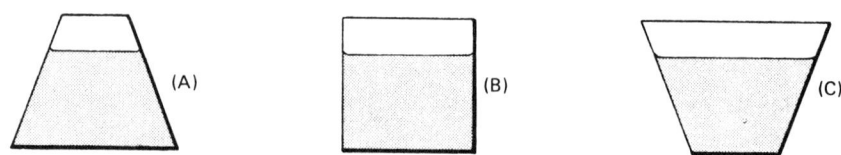

VESSELS A, B, AND C ARE FILLED TO THE SAME HEIGHT WITH WATER.

Fig. 18-7

1. The force on the bottom of A is (greater than) (less than) (equal to) the weight of the water.
2. The force on the bottom of B is (greater than) (less than) (equal to) the weight of the water.

3. The force on the bottom of C is (greater than) (less than) (equal to) the weight of the water.

4. The pressures on the bottom of A, B, and C are (not comparable) (the same) (different).

5. Pressure depends upon (area and density) (area, height, and density) (area and height).

6. The force on a container depends upon (area and height) (area and density) (area, height, and density).

Add the word or words needed to complete statements 7 to 10.

7. A liquid with a specific gravity of 0.70 _____ in water.

8. Carbon tetrachloride with a specific gravity of 1.5 _____ in water.

9. A substance with a density of 5.4 lb/ft^3 _____ in a liquid with a density of 62.4 lb/ft^3.

10. A liquid with a density of 45.4 g/cm^3 _____ in a liquid with a density of 27.2 g/cm^3.

11. What is the pressure at the bottom of a water column 50 ft high?

12. Explain why the flow of water in a faucet in the home decreases when another faucet is opened.

13. How is the pressure of a water system affected by (a) increasing the diameter of the standpipe to a water tower, and (b) by increasing the length or height of the standpipe?

14. State the advantages of applying Pascal's Law to a system of hydraulic brakes.

15. Find the weights, in the units indicated, for each of the following liquids, using an assumed weight for water of 8 lb/gal. Round off answers (where applicable) to one place.
    a. 5 gallons of cutting oil (specific gravity 0.70)
    b. 80 liters of gasoline (specific gravity 0.80) in kilograms (kg)
    c. 1 quart of carbon tetrachloride (specific gravity 1.50)
    d. 0.5 liter of sulphuric acid (specific gravity 1.82) in grams (g)

16. A tank 21 feet in diameter is filled with gasoline to a depth of ten feet. The gasoline weighs 0.78 times the weight of the water.
    a. Find the pressure on the bottom of the tank.
    b. Compute the force on the bottom.

17. a. Compute the force on the bottom of containers A through F and the pressure in each container per square foot. Use 62.4 lb/ft³ as the weight of water.
b. Convert the pound (lb) and lb/ft² values computed for force and pressure for A through F to equivalent newtons (N) and N·m², correct to two decimal places.

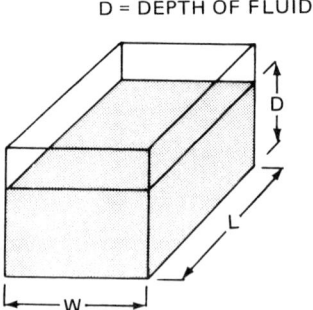

D = DEPTH OF FLUID

	Dimensions of Fluid			Specific Gravity	Customary Units		SI Metric Units	
	L	W	D		Force (lb)	Pressure (lb/ft)	Force (N)	Pressure (N/m²)
A	1 ft	1 ft	1 ft	1				
B	2 ft	2 ft	1 ft	1				
C	2 ft	2 ft	6 in	1				
D	2 ft	2 ft	2 ft	1				
E	10 ft	10 ft	10 ft	0.8				
F	10 ft	10 ft	5 ft	0.8				

Application of Fluid Pressure to Hydraulic Machines

1. The pistons of a hydraulic lift have an area ratio of 1 to 600. A force of 50 pounds is exerted on the small piston.
    a. Determine the force on the large piston.
    b. Determine the force distance through which the large piston moves when the small piston makes 25 four-inch strokes.
2. The small piston of a hydraulic jack has an area of 40 cm²; the large piston has an area of 80 cm². What force on the small piston is needed to raise a 1500-kg load?
3. The diameters of the pistons on a heavy hydraulic press are 1 1/2 inches and 24 inches. A force of 75 tons is needed to stamp and form a metal part. What effort must be applied to the small piston?
4. The piston of a hydraulic lift for automobiles is 35 cm in diameter. The device is operated by water from a city water system. What water pressure is necessary to lift a car if the total load is 1569 kg?

5. Determine the missing values for problems A, B, C, and D from the data given in the table for a simple hydraulic press. NOTE: The computed values for (ED) at C and D are the total distances that the small pistons move, and not the distance moved for each stroke.

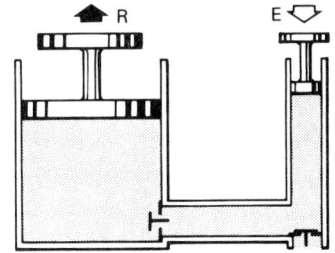

	E	Area of Small Piston	ED	R	Area of Large Piston	RD
A	10 lb	0.5 in²	5 in		5 in²	
B	50 lb	2 in²	5 in	500 lb		
C	45.4 kg	3.2 cm²		50 tons	3226 cm²	25.4 cm
D	34 kg	32.3 cm²		848.4 kg	806.5 cm²	10.16 cm

6. A hydraulic lift actuated by a lever is shown in figure 18-8. Compute the missing dimensions in the table for conditions A, B, C, and D.

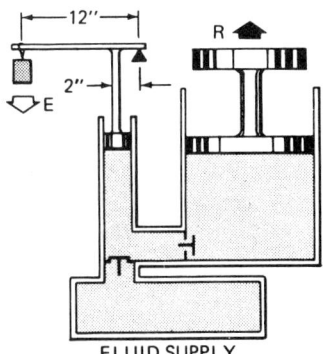

Fig. 18-8

	Effort Force (E)	Diameter Small Piston	Diameter Large Piston	Resistance Force (R)
A	10 lb	1 in	5 in	
B	25 lb		5 in	750 lb
C	45.4 kg	1.27 cm		5448 kg
D		5.08 cm	25.4 cm	1362 kg

7. The hydraulic lift in problem 6 is 80% efficient for conditions A and B. Determine the MA$_f$ for each condition.

8. Increase the effort distance to 50.8 cm for conditions C and D in problem 6. Using an efficiency of 75%, determine
   a. the mechanical advantage of force and
   b. the mechanical advantage of speed.

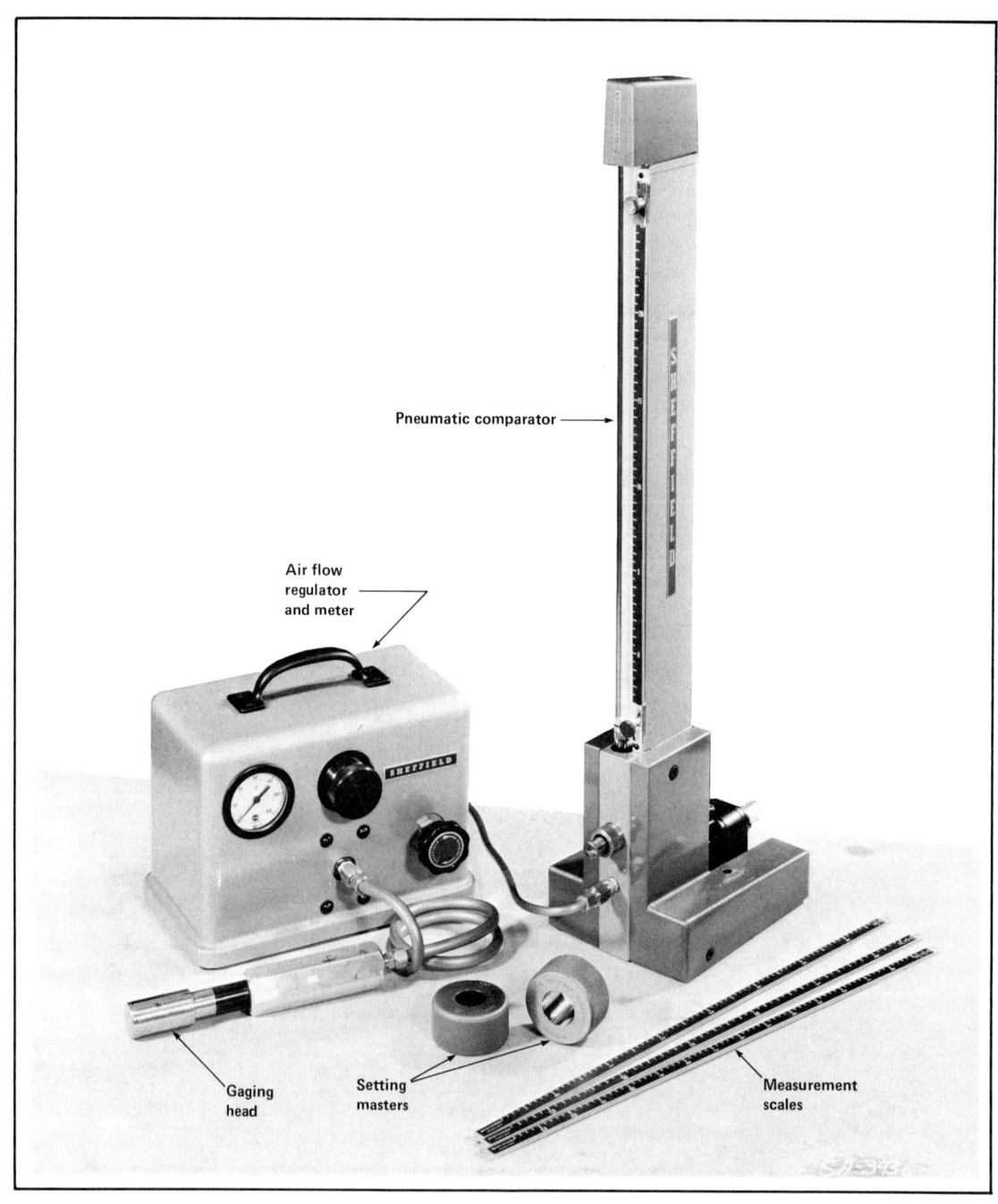

Flow-type pneumatic comparator for precision gaging. The size and geometry of holes depends upon air pressure and constant volume (Courtesy Automation & Measurement Div., Bendix Corp.)

Unit 19 Atmospheric Pressure: Principles and Applications

When the term *atmospheric pressure* is used, it refers to the pressure caused by an air ocean surrounding the earth's surface. Air has weight. Thus, air causes pressure on the surface of the earth because the force of gravity of the earth exerts a force on the atmosphere through a height of several miles. A cubic foot of air weighs about 1.2 ounces as compared to the same volume of water which weighs nearly 62.5 pounds. In other words, water weighs approximately 800 times as much as air. The total amount of air pressure exerted by this ocean of air will support a column of water at a maximum height of 34 feet. The air pressure required to support this column of water is equal to height × density: 34 ft × 62.5 lb/ft^3 = 2125 lb/ft^2. This value can be reduced to a pressure of 14.7 pounds per square inch or 1 033.2 grams per square centimeter.

■ MEASURING ATMOSPHERIC PRESSURE

The Barometer

While fluctuations in air pressure can be measured with a 34-foot column of water, such a measuring device is impractical. Evangelista Torricelli, an assistant to Galileo, perfected a simple device for measuring air pressure. In place of water, Torricelli used mercury (Hg) which has a specific gravity of 13.6. The rise of mercury in this device always reached about 30 inches (at sea level). As a result, a 30-inch column of mercury indicated the same atmospheric pressure as a 34-foot column of water. The mercury-filled measuring instrument became known as a *Torricelli barometer*. Torricelli's experiments showed that if, for any reason, the air pressure is lowered, the 30-inch height of mercury cannot be supported.

Pascal discovered that as a barometer is carried to higher altitudes, the mercury drops almost one inch for every 1000 feet of elevation, table 19-1. When a barometer is used for measuring altitude, it is usually called an *altimeter*.

Since air pressure varies with distance above the surface of the earth, it was necessary to establish a standard to provide a constant value from which to work. The standard pressure was selected to be a column of mercury 29.92 inches

Altitude, Ft Above Sea Level	Pressure	
	InHg	Lb/In^2
000 (Sea Level)	29.9	14.7
1000	28.9	14.1
2000	27.8	13.6
3000	26.8	13.2
4000	25.8	12.6
5000	24.9	12.2
6000	24.0	11.8
7000	23.1	11.3
8000	22.2	10.9
9000	21.4	10.5
10 000	20.5	10.1
15 000	16.9	8.3
20 000	13.8	6.8

Table 19-1 Atmospheric pressure versus altitude

or 76 centimeters high at sea level, figure 19-1. While pressure should be given in units of weight and square measure, it is common practice to refer to air pressure in terms of 29.52 inches, or 62 centimeters, or 752 millimeters. Each of these values refers to the height of a column of mercury which the atmospheric pressure will support.

In summary, the atmospheric pressure at sea level is equal to:

14.7 lb/in^2
29.92 inHg
76 cmHg or 760 mmHg
34 ft of water

In metric units, atmospheric pressure at sea level is

1.013 × 10^5 newtons/m^2; 101.33 kPa; or 1.033 2 kgf/cm^2

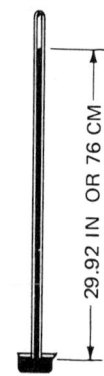

Fig. 19-1

The Aneroid Barometer

It is impractical to use a 30-inch column of mercury in many applications. A smaller and more compact device for measuring changes in air pressure is called the aneroid barometer.

The *aneroid barometer*, figure 19-2, consists of a corrugated metal expansion box from which the air is removed. Variations in air pressure cause the metal box to be compressed or returned to its original shape. A system of levers magnifies the movement of the corrugated metal box and transmits this movement to a pointer. The reading is indicated directly on a scale calibrated in inches of mercury.

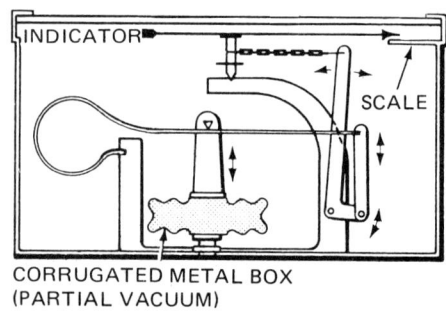

Fig. 19-2

If a continuous reading of pressure is required, a recording pen and chart and a timing mechanism are added to the barometer. In this manner, a permanent record is formed of changes in air pressure (or the pressure of other gases) for specific periods of time. This unit has described the simple mercury barometer and the mechanical aneroid barometer. However, many other types of barometers are available for measuring, recording, and controlling the pressure and volume of fluids.

■ ATMOSPHERIC PRESSURE AT WORK

Air pressure is important to the degree that it can be harnessed to do productive work. Before applications of air pressure can be made, there are several basic principles governing the action of air pressure that must be understood. Robert Boyle, an English scientist, found in 1662 that a high pressure is created by reducing the volume of a gas while maintaining a constant temperature. In addition, it was found that the pressure of an enclosed gas is inversely proportional to its volume, figure 19-3.

Boyle's Gas Laws

Figure 19-3 illustrates *Boyle's Law*. Starting at rest in (A), the volume of the gas in the cylinder is decreased to one-half at (B) and one-fourth the original volume at (C). For these same cylinders, the pressure is doubled at (B) and quadrupled at (C) (if the effects of heating due to compression are overcome).

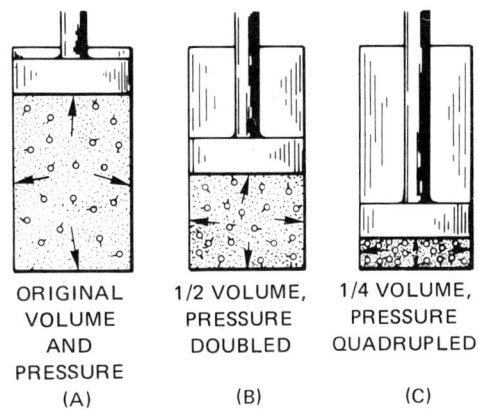

Fig. 19-3 Boyle's Gas Law

Boyle's Law can be explained on the basis of the principle which states that all matter is made up of tiny particles known as molecules. It is the bombardment of these molecules against the walls of a container that produces pressure. Thus, in figure 19-3A, the gas molecules exert a normal pressure against the walls of the container. When the volume of gas is compressed to half the original volume, the number of molecules per cubic inch doubles, figure 19-3B. In addition, the bombardment of molecules against this same piston area results in a pressure that is twice that of the original pressure. As the molecules are again compressed into one-fourth the original volume, the number of molecules and the pressure of these molecules per square inch is multiplied four times, figure 19-3C.

To this point, the unit has been concerned with the compression of gases. Figure 19-4 illustrates what takes place as the gas is expanded. If the original volume at (A) is increased to twice the original volume, as shown at (B), the molecules occupy twice as much space. The bombardment of molecules against the walls of the container is reduced to half the number of the original. As a result, the pressure is also reduced to one-half the pressure of (A). When the volume is expanded to four times the original volume, as shown at (C), the pressure is decreased to one-fourth the pressure at (A).

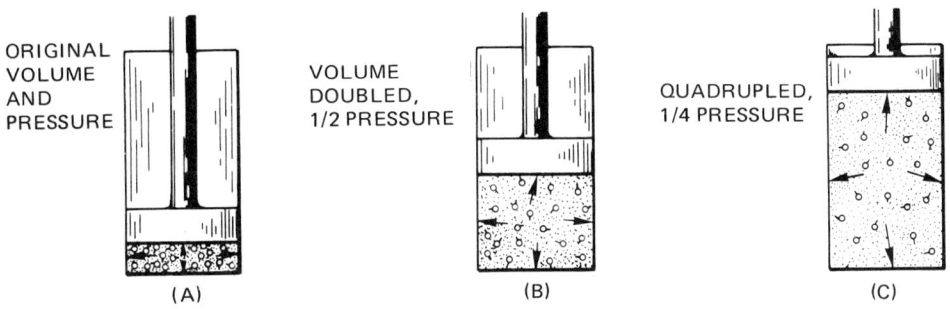

Fig. 19-4

Boyle's Law applies equally to low or high pressures and is expressed by a formula as follows:

$$P_1 \times V_1 = P_2 \times V_2$$

where P_1 = original pressure of a gas
V_1 = original volume of a gas
P_2 = pressure of a gas under a second set of conditions
V_2 = volume of a gas under a second set of conditions

For example, consider the case in which four cubic feet of a certain gas at a normal pressure of 15 pounds per square inch are compressed into a tank with a volume of one cubic foot. By substituting the given values in the formula, it is possible to determine the pressure that the tank must be able to withstand.

$$P_1 \times V_1 = P_2 \times V_2$$
$$15 \text{ lb/in}^2 \times 4 \text{ ft}^3 = P_2 \times 1 \text{ ft}^3$$
$$P_2 = 60 \text{ lb/in}^2$$

Thus, the pressure increases to compensate for the reduction in volume.

To convert a pressure in pounds per square inch to the SI metric equivalent in pascals (Pa), multiply by 6895. In the preceding example, 60 lb/in^2 × 6895 = 413 700 pascals (Pa). The writing of large pascal values is simplified by changing to kilopascals. 413 700 Pa = 413.7 kPa

Changing the Amount of Air

The simple air pump, figure 19-5, is the basis for more complex compressors which squeeze or force air under heavy pressures into closed vessels. The air pump consists of a piston moving up and down in a cylinder which also contains intake and outlet openings controlled by valves.

The upstroke of the piston causes the volume of air in the cylinder to increase and the pressure of the air to decrease. Normal atmospheric pressure forces air through the intake opening.

On the downstroke, the piston compresses the air in the cylinder. The intake valve is forced to close and the outlet valve opens. The air pump exhausts the air from one container and compresses it into another.

The Siphon

The action of a siphon is the result of air pressure. A *siphon* is a U-shaped tube with

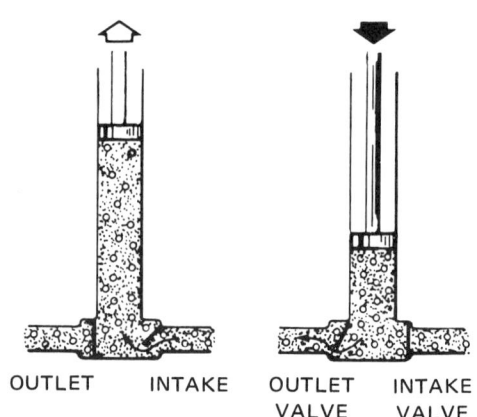

OUTLET   INTAKE   OUTLET   INTAKE
                                  VALVE   VALVE

Fig. 19-5

arms of unequal lengths, figure 19-6. This device is used to move liquids from a higher elevation to a lower point. The siphon is not self-starting, and must be filled with liquid to initiate the flow.

In figure 19-6, assume that water is to be siphoned from the container on the left. The atmospheric pressure on the surface of the water in each container is the same (approximately 15 lb/in^2). The pressure within the tube varies. Atmospheric pressure on the surface of the liquid maintains the liquid column in the left side of the tube. At a higher level, there is less pressure on the

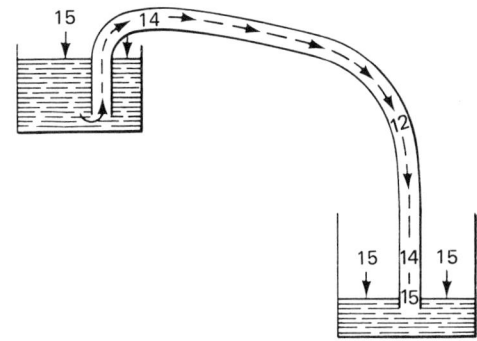

Fig. 19-6

liquid than at a lower level. Therefore, the pressure at the first (left) bend in the tube may be 14 lb/in^2 (0.98 kgf/cm^2). At the right (outlet) end, the pressure is atmospheric pressure (15 lb/in^2) (1.05 kgf/cm^2). Again, the pressure on the fluid decreases with the height above the outlet. The pressure difference from left to right (from 14 lb/in^2 to 12 lb/in^2 or 0.98 kgf/cm^2 to 0.84 kgf/cm^2) along the horizontal section of the tube causes the fluid to flow to the right (the fluid tends to flow toward a low-pressure area). Once the fluid reaches the area of the lowest pressure, 12 lb/in^2 (0.84 kgf/cm^2), the weight of the water causes it to fall into the lower container.

■ OTHER APPLICATIONS OF MECHANICAL DEVICES USING FLUIDS

Older Types of Lift and Force Pumps

The principles of air pressure can be combined with a pump to produce useful work to move fluids, figure 19-7. If the inlet side of a pump is connected to a liquid source, a pressure will force this liquid through the pump outlet. As the piston moves up, the volume of air in the cylinder increases and the pressure is lowered. Atmospheric pressure

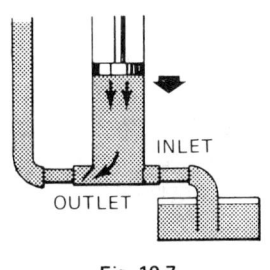

Fig. 19-7

on the liquid forces it into the pump cylinder. On the downstroke of the piston, the pressure exerted on the liquid by the piston closes the inlet valve, opens the outlet valve, and forces the liquid through the valve.

Water and other liquids can be raised by a simple lift pump, figure 19-8A. This type of pump depends on atmospheric pressure to start the upward movement of the liquid. The first few strokes of the piston usually reduce the pressure in the pipe until atmospheric pressure can force the water into the cylinder. The flow of liquid from a lift pump is intermittent and depends on the piston action. The lift pump cannot lift water from a well more than 34 feet (10.4 m) deep. The lift pump works by reducing air pressure and it has been shown that air pressure will not lift water beyond a height of 34 feet (10.4 m).

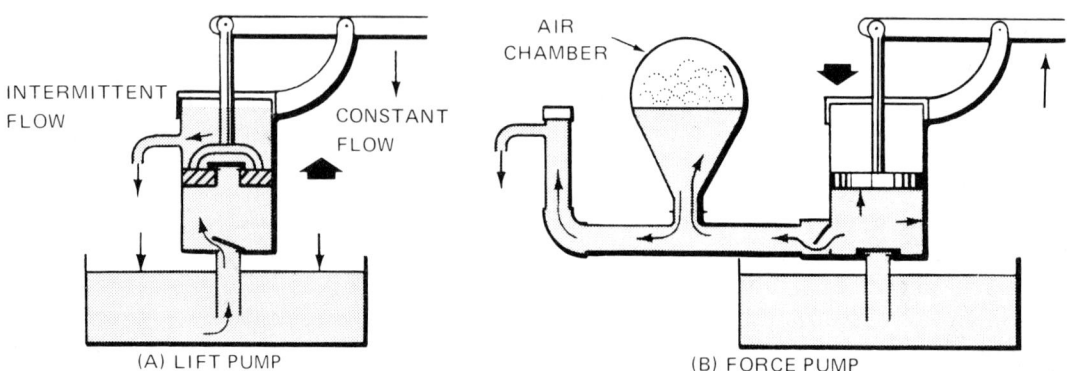

Fig. 19-8

When a continuous flow is needed, a force pump with an air chamber is used, figure 19-8B. The upstroke of the piston causes a reduction of the pressure inside the cylinder. Atmospheric pressure then forces the liquid into the cylinder. The downstroke of the piston forces the liquid through the outlet valve into an outlet pipe. The pressure of the downward stroke also compresses the air in the air chamber. On the upward stroke of the piston, the outlet valve closes and the compressed air continues to force the liquid through the outlet pipe to insure a continuous flow.

Figure 19-9 illustrates several types of fluid pumps commonly used. Figure 19-9A shows that an electric fan can create low- and high-pressure areas with air. The centrifugal pump, figure 19-9B, can pump large volumes of liquid under pressure. The rotary vane pump, figure 19-9C, is used when low pressures must be developed. The movement of the eccentric reduces the fluid pressure at the inlet side and forces the gas through the outlet side. Such a pump makes it possible to remove gas from a container until the pressure remaining is reduced to one-millionth part of one atmosphere. Another common type of fluid pump is the gear pump, figure 19-9D. This type of pump produces a flow at a constant rate and pressure.

■ USING FLUID PRESSURES IN INDUSTRY

Several of the numerous uses of air and liquid pressures were illustrated in this unit. The operation of many mechanical devices depends upon the use of air at greater than atmospheric pressure. As one example, energy can be transmitted great distances by piping air pressure which has been built up by a compressor. Air pressure is used in underwater construction such as tunnels and bridge piers to force water out of huge reinforced steel chambers called *caissons*.

All compressors, compressed air and liquid devices, and internal combustion and steam engines depend upon Boyle's Law. Other industrial processes using compressed air include welding and cutting, air conditioning, and spraying.

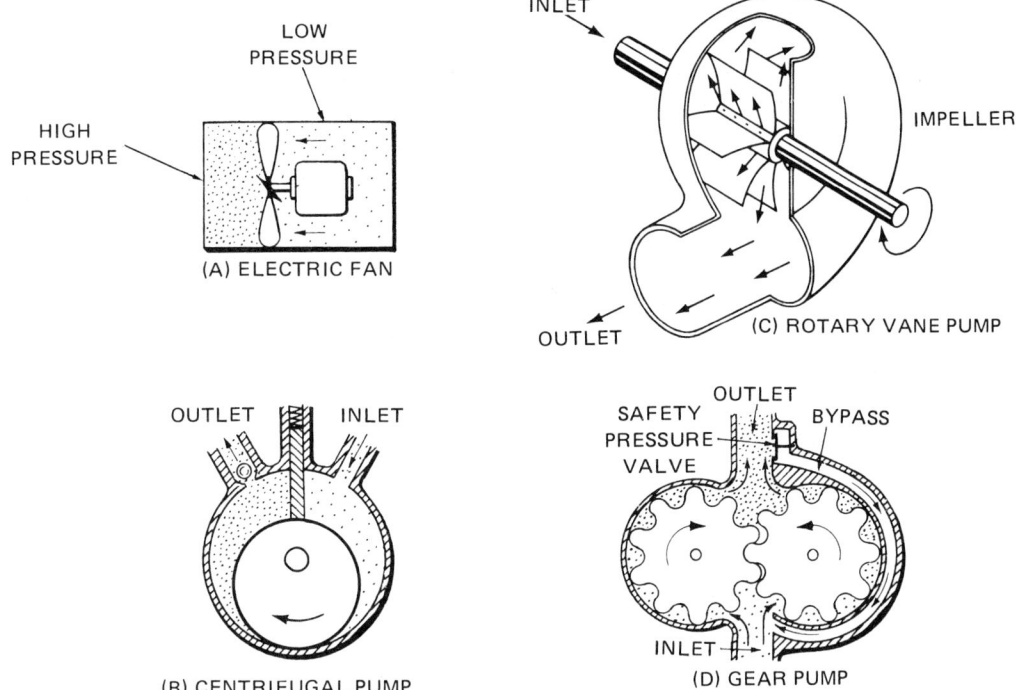

Fig. 19-9

SUMMARY

- The atmosphere exerts pressure equally in all directions.

- Atmospheric pressure at sea level can support a column of water 34 feet high or a column of mercury 30 inches high.

- The normal atmospheric pressure at sea level is 14.7 lb/in^2, or 29.92 inHg, or 76 cmHg, or 1.033 2 kgf/cm^2, or 101.33 kPa.

- Atmospheric pressure decreases almost one inch for each rise of 1000 feet, or 2.54 cm for each rise of 304.8 meters.

- The Torricelli barometer is an instrument which measures the height to which atmospheric pressure, under varying conditions, supports a column of mercury.

- The aneroid barometer is a practical measuring device for determining and recording air pressure.

- Boyle's Laws, which apply to decreasing or increasing pressure (when the temperature remains constant), state:
  — A high pressure can be created by reducing the volume of a gas; a low pressure can be obtained by increasing the volume of a gas.

  $$P_1 \times V_1 = P_2 \times V_2$$

- The siphon is a U-shaped tube, or similar device, used to move liquids from one level to a lower level.
- Pumps are used to lift or force fluids through a mechanism or a piping system. The pump action can result in an intermittent or a constant flow. Such pumps operate on the principles of Boyle's Laws.
- Mechanical devices such as compressors, fans, rotary vane and gear pumps, pneumatic tools, and numerous other machines produce either high or low pressures and volumes with fluids. All of these devices depend upon Boyle's Laws.

ASSIGNMENT UNIT 19 ATMOSPHERIC PRESSURE: PRINCIPLES AND APPLICATIONS

■ PRACTICAL PROBLEMS IN APPLYING ATMOSPHERIC PRESSURE

Measuring Atmospheric Pressure

Select the correct word or phrase to complete statements 1 to 7.

1. The weight of a cubic foot of air is (62.5 lb/ft^3) (14.7 lb) (1.2 oz).
2. The standard atmospheric pressure at sea level is (101.33 kPa) (101.33 Pa) (101.33 kg/cm^2).
3. Atmospheric pressure at sea level supports a column of mercury (30 in) (34 ft) (62.5 in) high.
4. Atmospheric pressure at sea level supports a column of water (76.2 cm) (10.37 m) (100 cm) high.
5. A change of 1000 feet in elevation causes (a change of 2 inches) (a change of 1 inch) (no change) in the height of a column of mercury.
6. The (Torricelli) (aneroid) barometer is a compact device for measuring and recording changes in pressure.
7. Boyle's Laws (are) (are not) affected by a raising or lowering of temperature.
8. State the relationship between the density and the pressure of gases.
9. As a weather balloon or other balloon rises, does the pressure (a) in the balloon increase, and (b) against the balloon increase? Explain.
10. Compare (a) the air pressure at both ends of a siphon, and (b) the pressure on the liquid in both containers.

11. List two commercial applications of the siphon and two household uses.
12. Compare the advantages of the aneroid barometer to those of the mercury barometer in terms of (a) sensitivity, (b) accuracy, and (c) convenience in handling, reading, and recording.
13. From the data given in the table determine (a) the column height of fluids A and B which will be supported by atmospheric pressure at sea level, and (b) the specific gravity of fluids C and D.

	Specific Gravity	Column Height
A	6.8	
B	0.82	
C		20.74 m
D		7.4 m

Atmospheric Pressure at Work

Add a word or phrase to complete statements 1 to 12.

1. The pressure of gases in a confined space is caused by the bombardment of _____ against the walls of the container.
2. When the original volume of a gas is compressed, the _____ increases.
3. Any change in pressure on a confined gas produces a change in _____.
4. When the temperature of a gas is constant, doubling the pressure _____ the volume.
5. Without any change in temperature, decreasing the pressure 50% _____ the volume _____.
6. If the temperature remains constant, a reduction of the volume of a gas produces a corresponding _____ in pressure.
7. Boyle's Law, stated mathematically is: _____.
8. A _____ is a simple tube device for moving liquids from one level to a lower level.
9. The siphoning action _____ when the liquid level in one container is the same height as the level of liquid in the long arm of the siphon.
10. For a siphon, the maximum height to which water can be raised by atmospheric pressure is _____.
11. The siphon works on the principle of a difference in _____ between the pressure inside the tube and the pressure on the surface of the fluid.
12. In addition to the diameter of the tube, the rate of flow by siphoning depends upon the _____ between the two liquid levels.

13. Explain how an air pump works. Include a simple sketch with the important parts labeled.
14. On what principle does a siphon operate? Explain and include a sketch.
15. Using a simple sketch, explain how a vacuum cleaner works.
16. Compute the height to which fluids (A) and (B) may be raised for the values given in the table. Assume an atmospheric pressure of 15 lb/in². Round off answers to one decimal point.

	Pressure	Height — Fluid A (SG 0.8)	Height — Fluid B (SG 1.5)
A	5 lb/in²	ft	ft
B	10 lb/in²	ft	ft
C	15 lb/in²	ft	ft
D	1.758 kg/cm²	m	m
E	2.812 kgf/m²	m	m

17. Graph the values determined in question 16 for volume (height) and pressure. Use a horizontal scale for the volume of 1/2 in = 10 ft (height of column) and a vertical scale of 1/2 in = 5-lb pressure. Convert the meter values to feet and plot these values for D and E on the graph.
18. Refer to the graph prepared in question 17 and (a) tell what happens when the pressure is increased, and (b) determine the volume of fluid (A) and fluid (B) for pressures of 20, 30, and 35 lb/in².

Applications with Mechanical Devices

1. Match each type of pump or fan in Column I with the appropriate use given in Column II.

    Column I
    a. Centrifugal fluid pump
    b. Electric fan
    c. Rotary vane pump
    d. Gear pump

    Column II
    1. Exhausts gases from tubes to 1/1 000 000 th part of an atmosphere.
    2. Increases or decreases atmospheric pressure by the action of straight-line motion pistons.
    3. Pumps large volumes of fluids at a fairly constant rate.
    4. Creates high or low pressure areas with air.
    5. Pumps fluids at a constant rate and pressure.

Select the correct word or phrase to complete statements 2 to 5.

2. The simple lift pump used to raise fluids produces (a constant) (an intermittent) flow.
3. The force pump (which uses the force of a piston and an air chamber) produces (a constant) (an intermittent) flow.

4. (Air pressure) (Water pressure) is used to prevent the flow of water into an underwater caisson.

5. Acetylene gas is supplied at a controlled volume and pressure from cylinders. To do this, the original volume (is reduced at atmospheric pressure) (is reduced under greatly increased pressure) and stored in a cylinder.

6. The 8.89-cm diameter piston of a single-cylinder engine compresses a fuel mixture to one-sixth of its original volume of 360.5 cm^3. Determine the final pressure on the piston at the end of the compression stroke.

7. The table furnishes data on the volume and pressure for gases A, B, C, and D. The volume consumed for each hour of operation is also given. Assuming the atmospheric pressure is 15 lb/in^2, determine the volume of gas that can be used by reducing the pressure to atmospheric pressure. Determine the number of hours the fuel can be used as the pressure is decreased to atmospheric pressure.

	Volume (In^3)	Pressure (Lb/In^2)	Vol. when P is Reduced to 15 Lb/In^2	Hourly Consumed Vol. (In^3)	Hours Supply of Gas
A	1500	45		1000	
B	1500	165		1000	
C	3000	165		750	
D	3000	457.5		500	

Unit 20 Fluid Power: Principles and Applications

Fluids under pressure are widely used in systems that transmit and control power. When the fluid is air or any other gas, the system is identified as a *pneumatic* system. A system that uses oil or another type of liquid is known as a *hydraulic* system. These two types of systems are sometimes combined into a common system called *fluid power*.

■ PRESSURE, FORCE, AND VOLUME RELATIONSHIPS

Fluid power applications depend upon the principles of pressure to exert a force or torque and the principle of *flow*. The output of the system in terms of the speed and amount of motion produced, is related directly to flow.

When a pressure is applied to a gas, the molecules of the gas are pushed closer together and are compressed into a smaller space. The greater the force pushing the molecules together, the greater is the effort of the molecules to move apart. As the pressure on a gas in a closed container increases, the volume decreases, figure 20-1. Conversely, as the volume of a gas is increased, the pressure decreases.

Figure 20-1 reviews Boyle's Law. Part (A) shows a cubic foot of gas under ten pounds or 44.5 newtons of pressure. The volume is decreased at (B) to one-half cubic foot under 20 pounds (89 newtons) of pressure. At (C) the volume is reduced further to one-tenth cubic foot under 100 pounds (445 newtons) of pressure. Using the formula $P_1 \times V_1 = P_2 \times V_2$, the volume of the gas (V_2) under 20 lb of pressure is 0.5 cu ft: $(10) \times (1) = 20 (V_2)$. The same formula can be applied to (C) with the result that, $V_3 = 0.1$ cu ft.

In contrast to gases, a liquid resists efforts to be compressed and reduced in volume. The volume of the liquid shown in figure 20-1, for practical purposes, is the same at (A), (B), and (C). However, the same force can be transmitted by either the liquid or the compressed air or gas.

The effect of heat on fluids is another important consideration in the control and transmission of power. The effects of temperature and heat energy are covered in later units. However, at this point, it should be noted that fluid power systems must be

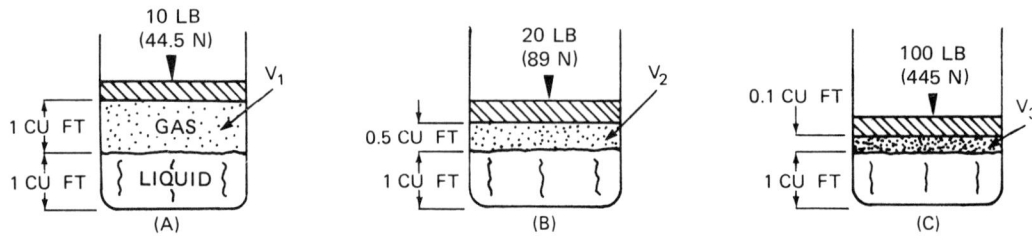

Fig. 20-1 Compressibility of fluids

designed to compensate for the expansion of fluids due to changes in temperature resulting from pressure and flow factors.

Transmitting Power

Power can be transmitted efficiently through a fluid system. The pressure and direction of the force may be changed by the design and length of the transmission line or pipe through which the fluid flows. The change of direction does not affect the transfer of force, figure 20-2.

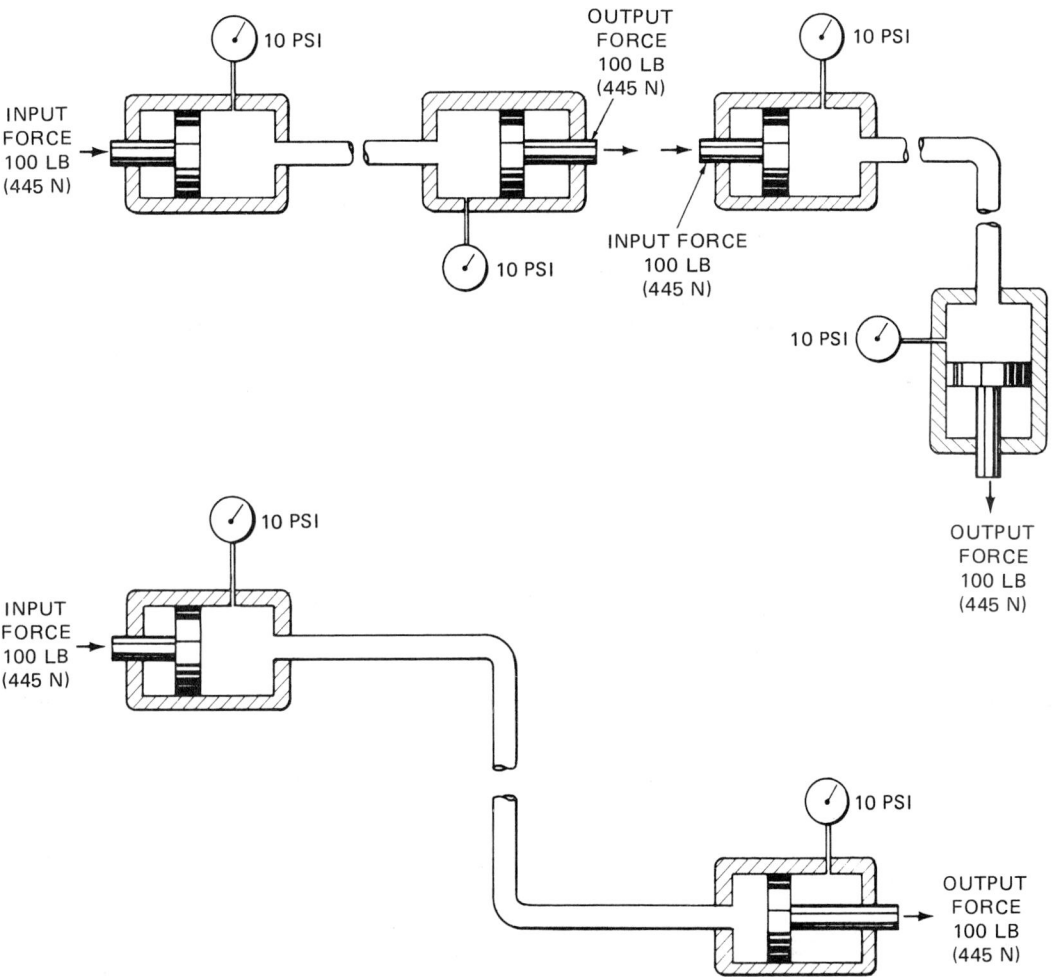

Fig. 20-2 Transfer of force is not affected by a change of direction

176 Mechanics, Machines, and Wave Motion

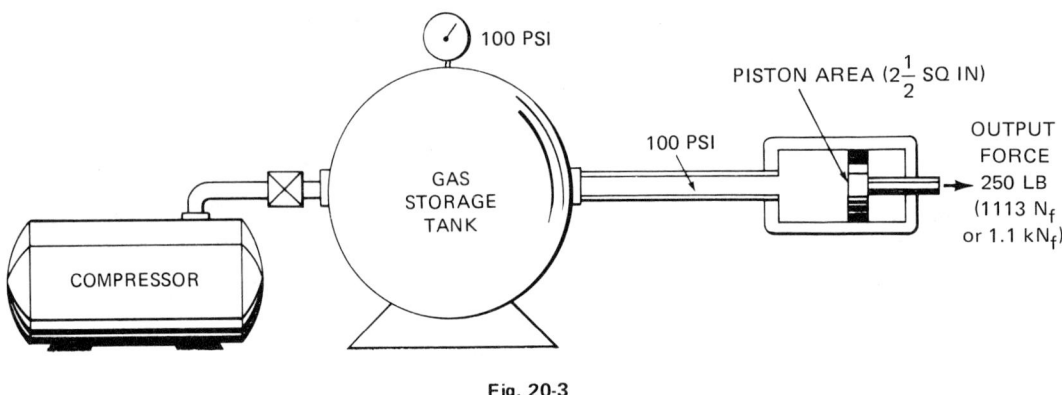

Fig. 20-3

Transmission of Force as a Pressure

Hydraulic systems have great flexibility. Forces can be applied over long distances by the use of fluid power. In situations where locations are inaccessible to standard mechanical linkages, the force can be transmitted by the addition of piping and liquid. All power systems are governed by the principle that a gain in force results in a loss of distance. The output force of a hydraulic system can be changed by changing the area of the surface upon which the force acts.

A pneumatic system normally uses a compressor to develop the pressure in a gas. A storage tank holds the gas at this pressure. The pressure is then transmitted through the system to the point of application, figure 20-3. The 100-psi pressure in the system can deliver a force of 250 pounds (1113 N or 1.1 kN) on a piston area of 2 1/2 square inches or 16.13 square centimeters.

■ OPPOSITION (RESISTANCE) TO FLOW

Opposition is present whenever a force acts to change a motion or flow and another force is present to oppose the original force and limit the change of motion. In electrical circuits, the opposition is called *resistance;* in magnetic circuits, it is known as *reluctance;* in mechanical systems, *friction;* and in fluids, *fluid resistance.*

Fluid resistance limits the change of motion or flow and transforms some of the energy into heat (thermal) energy. All fluids have varying degrees of viscosity. Viscosity accounts for the internal friction (drag) of one part of a fluid on a neighboring part of the fluid in motion. The fluid also exhibits different types of motion. The flow may be *laminar* (steady), or it may be classified as *turbulent flow,* figure 20-4. Turbulent

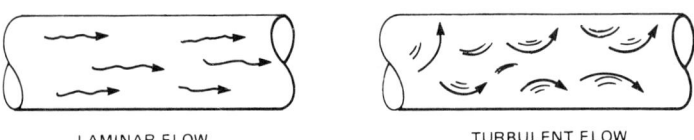

Fig. 20-4

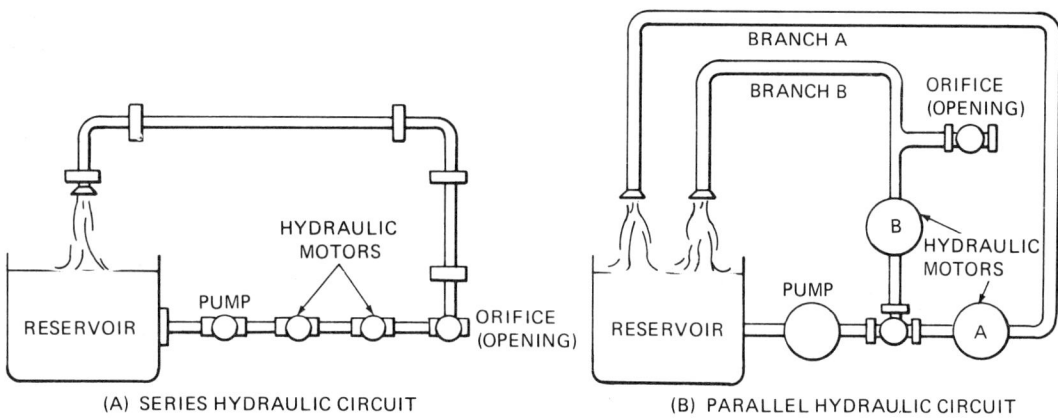

Fig. 20-5

flow has whirls and eddies due to the rapid motion or to obstructions in the flow path. Friction losses are minimized in laminar flow where the layers of liquid move smoothly over one another.

There are two basic types of fluid circuits. In a *series circuit*, figure 20-5(A), all of the fluid flows through each component in the system. In a *parallel circuit*, figure 20-5(B), the fluid flows through branches which are designed to provide the same or a different rate of flow and/or pressure. Part of the fluid passes through one set of components and, at the same time, another part of the fluid passes through other sets of components. Series and parallel circuits may be combined into a *series-parallel network*.

Total Resistance of Fluid Systems

Energy is required to maintain the flow or motion of fluids and to overcome resistance within the system. As the length of pipe or the number of openings (orifices) within the system increases, the resistance also increases. The total resistance (R_t) is equal to the initial resistance (R_1) plus the resistance (R_2) of any additional pipe or of a second orifice, and so on until the last resistance in the system is accounted for.

$$R_t = R_1 + R_2 + \ldots + R_n$$

where R_n is the last resistance in the system.

Resistance of a Series Circuit

As the pressure is increased in a series circuit, the quantity of fluid per unit time which passes through the circuit is increased. This quantity of fluid per unit time is the *fluid flow rate*. Any opening (orifice) placed in the circuit will decrease the flow rate for the same pressure.

The total resistance (R_t) of a series circuit is equal to the pressure difference (ΔP) across the circuit divided by the quantity of fluid (Q) transported per unit of time in the system.

$$R_t = \frac{\Delta P}{Q}$$

Resistance of a Parallel Circuit

In a parallel circuit, the total resistance can be calculated by using one of two formulas. When the resistance in each branch of a parallel circuit is equal, the total resistance (R_t) is equal to the resistance (R) divided by the number of branches (N).

$$R_t = \frac{R}{N}$$

If the resistance of the two branches is not equal, the total resistance is obtained using the equation:

$$R_t = \frac{(R_1) \times (R_2)}{R_1 + R_2}$$

The total resistance of a parallel circuit with more than two resistances can be calculated using the equation:

$$R_t = \frac{1}{\frac{1}{R_1} + \frac{1}{R_2} + \frac{1}{R_3} + \ldots + \frac{1}{R_n}}$$

Series-parallel Fluid Flow

Some fluid flow systems require a combination of any number of series and parallel circuits. Pressure differences are necessary to maintain the fluid in motion. To compute the resistance of a combination series-parallel system, an easy method is to consider the system as a simple series circuit. The total resistance of the system then is the sum of the resistances of the equivalent series component parts. The total resistance is expressed in psi-min/ga. The flow rate may be expressed in gpm (gallons per minute) or L/min (liters per minute); or cu ft/min (cubic feet per minute) or m^3/min (cubic meters per minute).

■ STATIC STORAGE — POTENTIAL AND KINETIC ENERGY

A fluid system requires a storage tank and reservoir. This reservoir serves three prime functions in that it (1) maintains a constant pressure head, (2) allows heat to be removed, and (3) stores energy. The stored energy is *potential energy* when the fluid has the ability to do work because of its position with respect to other bodies. The potential energy (PE) of a body (in this case the fluid) is equal to the weight of the body (W) multiplied by the height (H) the body is lifted.

$$PE = (W) \times (H)$$

Since weight (W) = mass (M) times the acceleration due to gravity (g),

$$PE = (M \times g) \times (H).$$

When the mass is released, the stored-up potential energy becomes the energy of motion or *kinetic energy*. The velocity of the mass when the body reaches the lowest position is the same as the speed of a body falling from rest from height (H). Torricelli developed a theorem to express this condition.

$$V = \sqrt{2gH}$$

This expression says that the speed at which the liquid is discharged is the same as the speed of a body falling from rest from height H.

The kinetic energy (KE) is equal to one-half the mass (M) multiplied by the square of the velocity (V).

$$KE = 1/2 M(V^2)$$

The flow rate at which the liquid leaves the valve or orifice in the storage tank is given by the following equation:

$$Q = (V)(A) = A\sqrt{2gH}$$

where Q = flow rate, in cu ft/s
V = velocity, in ft/s
A = effective area of the opening, in ft^2
g = 32 ft/s^2
H = height from top of liquid to orifice, in ft

Several other factors are important in any consideration of fluid systems. For example, thermal resistance may be present as a result of temperature differences which cause heat flow from one body to another. There is a time factor to consider since time is required to flow a charge or mass through the system. The units for thermal flow are expressed in British thermal units (Btu) per second or minute.

The resistance of the system also depends upon the condition of the components. Corroded pipes and valves require an increase in pressure difference to maintain a given flow.

Time Constants

The term *time constant* is used to indicate the period of time required to complete the fluid movement in a system when the transfer continues at its initial rate. Any change from one pressure level to another requires time. If a pneumatic system is subjected to a sudden change of input, there is a time lapse before the system responds completely to the change. A system action of this type is known as a *transient response* condition. A *steady-state* condition means that the system has reached equilibrium and is constant.

Widespread Application of Fluid Power in Industry

In industry, pneumatic systems are classified as power systems and control systems. Pneumatic power systems are used for light loads and in situations where speed is important. Hydraulic systems are used with heavy loads where speed is of secondary importance. Pneumatic lifts, machine clamping devices, and pneumatic punches are typical applications of pneumatic power systems.

Pneumatic controls are widely used because of the advantages they have over electrical controls and other types of control systems. Air valves are fireproof and can be used in explosive areas. They do not overheat and burn out, and operate over long periods of time without malfunction.

Logic control methods are another application of pneumatic controls. Pneumatic logic control takes place without moving parts by using fluids in motion. This field is known as *fluidics. Fluidic control devices* have the capacity to determine an alternate course of action when the directed course proves to be undesirable. Some advantages of fluidic control devices include their compact size, low cost, and long operational life. However, the response time for fluidic controls is usually slower than that of a sophisticated electronic control system.

Hydraulic systems are used to operate forming and cutting presses, machine cutters, mills, and for other heavy-duty operations. The principles of fluids at rest and in motion are applied to motor vehicle components such as the brake system. This system transmits motion, transmits pressure to develop friction between the brake shoes and drums, and multiplies the force applied at the brake pedal to the pistons which actuate the brake shoes.

Transmission units, hydraulic jacks, hoists, and landing gear on aircraft are other typical examples of hydraulics at work. The automobile carburetor is an excellent example of fluids in motion and the application of Bernoulli's principle.

SUMMARY

- The term fluid power is applied to pneumatic (air/gas) systems and hydraulic (liquid) systems.
- Fluid power systems are used to transmit and control power, to apply force in places that are inaccessible to mechanical linkages, and in fluidic devices that regulate courses of action for automated mechanisms.
- The output force of a fluid power system can be changed by changing the surface area over which the force acts.
- Opposition (resistance) to flow results from forces that act to change the motion or flow. Fluid viscosity, the nature of flow, and the type of system are some factors which provide the opposition to flow.

- The total resistance of a fluid power system is equal to the initial resistance plus the resistances of additional pipes, orifices, valves, and other components.

$$R_t = R_1 + R_2 + \ldots + R_n$$

- The total resistance of a series circuit is equal to the pressure difference across the circuit divided by the quantity of fluid transported per unit of time.

$$R_t = \frac{\Delta P}{Q}$$

- The total resistance of a parallel circuit can be determined as follows:
  - For two equal branches, total resistance equals the resistance divided by the number of branches:

$$R_t = \frac{R}{N}$$

  - For two unequal branches, the total resistance is:

$$R_t = \frac{(R_1) \times (R_2)}{(R_1) + (R_2)}$$

- The total resistance, expressed in psi-min/gal, is the sum of the equivalent series component parts.
- The potential energy of a fluid body equals the weight of the body times the height the body is lifted. $PE = (W)(H)$.
- The kinetic energy that a body possesses due to motion equals $1/2 M(V)^2$.
- The velocity of the mass at the lowest position equals the square root of $2(gH)$: $V = \sqrt{2gH}$.
- The flow rate of the liquid at the orifice in a storage tank equals the effective area of the opening in ft^2 multiplied by the square root of 2 times g ($32 \, ft/s^2$), times the height (H) from the top of the liquid to the orifice, in feet.

$$Q = A\sqrt{2gH}$$

- A time constant in a fluid system refers to the time required to complete the fluid movement process at the initial rate.
- Pneumatic and hydraulic systems are widely used in industry, agriculture, and commerce. In addition, these systems control the operation of many mechanisms that serve individuals such as the automobile, other forms of transportation, and computer devices.

ASSIGNMENT UNIT 20 FLUID POWER: PRINCIPLES AND APPLICATIONS

■ PRACTICAL PROBLEMS WITH FLUID POWER CIRCUITS

Opposition (Resistance) in a Fluid Power Circuit

1. Calculate the resistance values and pressure drop across a series fluid circuit that has a rate of flow of six gallons per minute and components A, B, and C.

   Note: a. Compute the resistance for the number of components in the circuit and the total resistance (R_t) of all components.
   b. Determine the pressure drop resulting from each set of components and the total pressure drop across the circuit.

	Component	Resistance		Pressure Drop
		Unit Value	Total	(Psi-min/gal)
A	50-ft hydraulic hose, 1/4-in diam.	1.67	83.50	
B	Quick disconnect couplers (3)	4.9	14.70	
C	Flowmeters (2)	0.8	1.60	
			R_t:	Total P:

2. Record the unit resistance values for components A, B, and C. The components are part of a series fluid system which maintains a pressure of 240 psi.

   a. Compute the resistance for each set of components and the total of all components.
   b. Calculate the rate of flow for the system.

	Component	Resistance	
		Unit Value	Total for Components
A	12.2-m hydraulic hose, 0.635-cm diam.	1.67	
B	Pressure gages (4)	5.1	
C	Quick disconnect couplers (8)	4.9	
	Pressure drop:	Rate of Flow:	

3. Determine the resistance of a fluid power system which has a flow rate of 1.59 L/min under a pressure of 516 psi.

Fluid Energy

1. A fluid flows from a four-inch outlet of a storage tank at a speed of 18 ft/s.
   a. Determine the flow rate of the fluid.
   b. Convert the flow rate in cfs to m^3/s.

2. A fluid power system has an orifice which passes fluid at the rate of 9.6 ft/s and has a fluid flow rate of 300 gallons per minute. Compute the diameter of the outlet opening.
3. Determine the kinetic energy of 400 ft^3 of water in a storage tank when the fluid passes an outlet valve at 12.2 ft/s.

Unit 21 Wave Motion: Transfer of Energy

Energy can be transferred from one place to another by mechanical and other physical means of moving material. Energy can also be transferred by nonphysical transport. This unit deals with wave motion principles and phenomena which are developed further in later units through applications to heat, sound and light energy, and to selected electromagnetic waves.

■ MECHANICAL WAVES

Wave motion refers to the transfer and transport of energy by means of a disturbance in a medium. A *mechanical wave* must be generated by a mechanical source and must have a material medium through which it can move. In other words, a mechanical wave is considered to be a physical disturbance in an elastic medium.

As an example of a mechanical wave, consider the wave that is created when an object is dropped into a body of water. The disturbance spreads in the water as a series of concentric circles. That is, successive water particles transfer the energy from the point of contact in the fluid to an object that may be floating in the fluid. Thus, the waves transmit energy from one place to another through the motion created by a change in the medium, figure 21-1.

The student must recognize that disturbances (energy) can be transmitted physically on mechanical waves or without a physical medium. For example, energy is transmitted in a nonphysical medium when it is propagated by electromagnetic waves caused by electrical and magnetic disturbances.

■ CHARACTERISTICS OF WAVES

Types of Waves

There are two major types of waves: transverse and longitudinal. In a *transverse wave* the vibration of the individual particles of the medium is perpendicular to the

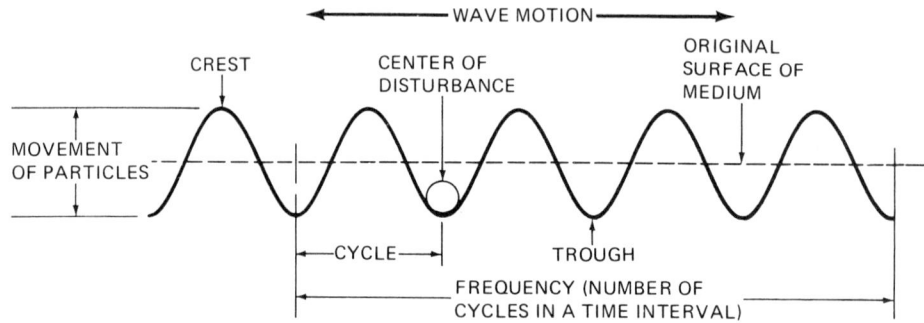

Fig. 21-1 Wave motion produced by a disturbance of the medium

direction of wave propagation. In the example of an object dropped into water, the resulting waves travel away from the point of impact and the particles of water move up and down at right angles to the line of travel of the waves, figure 21-2.

Thus, a single disturbance, called a *pulse*, is sent across the medium. As the individual particles of water move up and down, the disturbance moves to the left or right with a velocity (v).

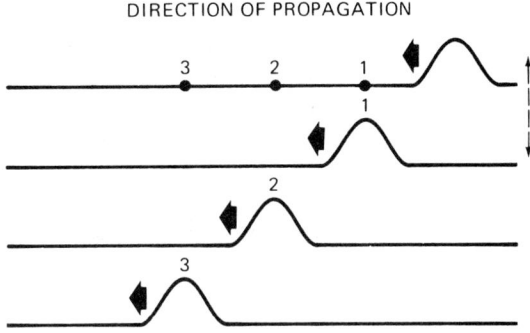

Fig. 21-2 Characteristics of a transverse wave

In a *longitudinal wave*, the vibration of the individual particles moves in a determined direction which is parallel to the direction of wave propagation. A coiled spring can be used to demonstrate the action of a longitudinal wave. The spring windings at the left of figure 21-3(A) are close together (condensed). When part of a wave is compressed in this manner, it is said that *condensation* is taking place. When the force causing the condensation is removed, a *condensation pulse* is created and propagated through the length of the spring, as shown in figures 21-3(B) and (C). The wave is a longitudinal wave because the material particles (the coils of the spring) are displaced in the direction of the disturbance.

If the coils in the spring are forced apart, figure 21-3(D), a *rarefaction* is formed. Again, when the disturbance force is removed, a longitudinal *rarefaction pulse* is propagated along the spring, as shown at (E) and (F) of figure 21-3.

A longitudinal wave generally consists of a series of condensations and rarefactions which move in a specified direction.

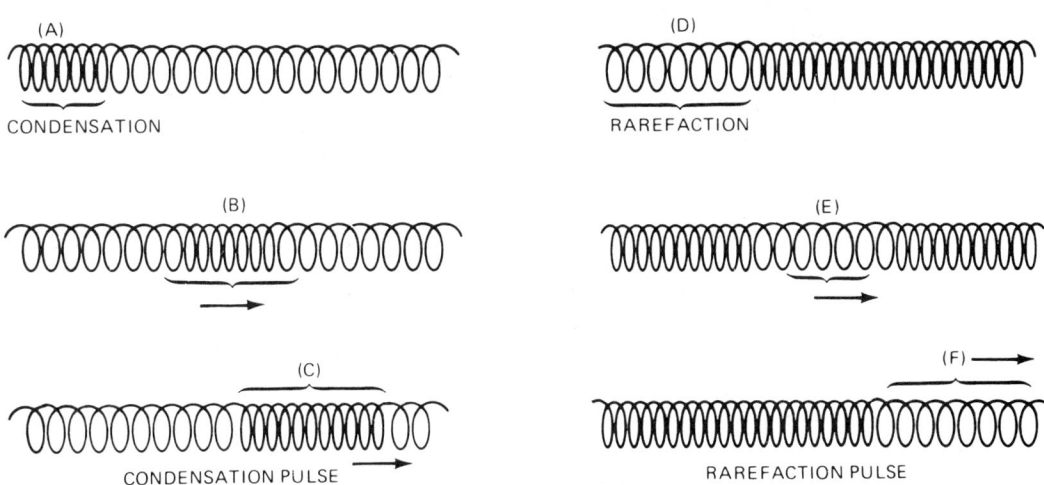

Fig. 21-3 Longitudinal motion

Pulses

A string can be used to demonstrate the simplest kind of wave phenomenon. When a completely flexible and uniform string is used, the pulse keeps the same shape as it moves along the stretched string with a constant velocity (v). The velocity depends upon the size of the string and the amount of tension caused by the stretching of the string.

The size of the string means the mass per unit length. A string with a large mass causes the pulse being transmitted to have a low velocity. The inertia of each segment of the string is high; thus, the string responds slowly to the forces acting upon it. The pulse velocity is high when the string is tightly stretched. The higher the tension of the string, the greater is the tendency for the string to straighten.

$$\text{velocity} = \sqrt{\frac{\text{tension}}{\text{mass/length}}}$$

$$v = \sqrt{\frac{T}{m/l}}$$

In a traveling pulse, the forward portion of the pulse moves upward and the back portion of the pulse moves downward, figure 21-4. If the end of the string is securely fastened, the arrival of an upward pulse at the attached end of the string exerts an upward force on the support, figure 21-5.

According to Newton's Third Law of Motion, the support exerts an equal and opposite reaction force on the fixed end of the string. As a result of this reaction force, an inverted pulse is propagated in the opposite direction, figure 21-5.

The inverted pulse has the same shape as the original pulse and moves back along the string to the free end. In summary, when an upward pulse is reflected at the fixed end, it becomes a downward pulse; similarly, a downward pulse becomes an upward pulse.

If the end of the string is free to move, figure 21-6, the pulse first moves upward on the string and then down again. While the direction of the pulse is reversed upon reflection, the reflected pulse has the same shape as the original pulse. If the tension on the string is

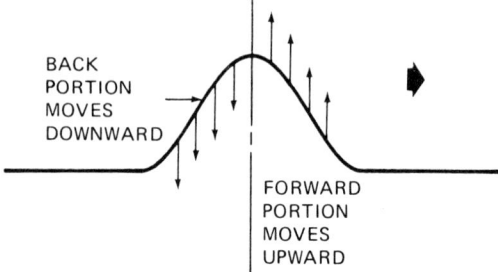

Fig. 21-4

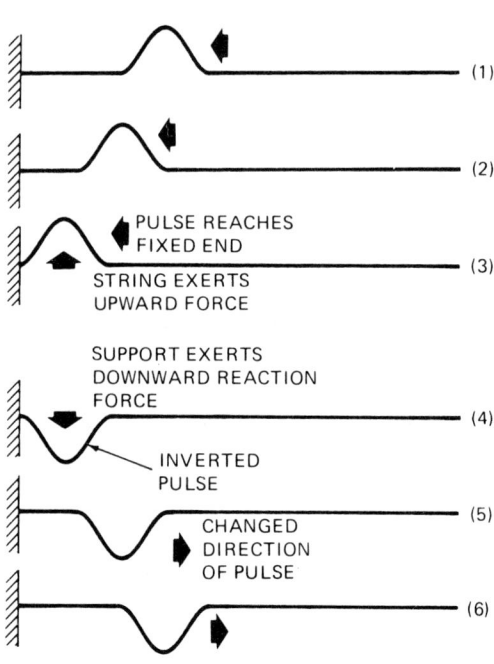

Fig. 21-5

between that of a completely taut string and a completely free string, the pulse disappears and is not reflected when it reaches the opposite end.

Pulses Transmitted Between Media of Low and High Mass Per Unit Length

It has been shown that pulses can travel in a material of uniform mass per unit length. Pulses can also travel through materials having different masses per unit length. In figure 21-7, a light string of low mass per unit length is attached to a heavier string that is rigidly fastened. A pulse, generated by the up-and-down motion of the free end of the string, passes from the light string to the heavy string at their junction. The pulse is transmitted in the same direction to the heavy string. The greater inertia of the heavy string causes a reactive force in the opposite direction. This reactive force produces an *inverted reflected pulse*.

When the pulse is transmitted from the heavy string to the light string, the inertia of the light string allows it to follow the pulse movement of the heavy string. However, in this case, the pulse velocity in the light string is greater and the length of the pulse is longer. The reflected pulse remains right side up. The preceding demonstration of the reflection and transmission of a pulse at a junction between different media also applies to other types of waves.

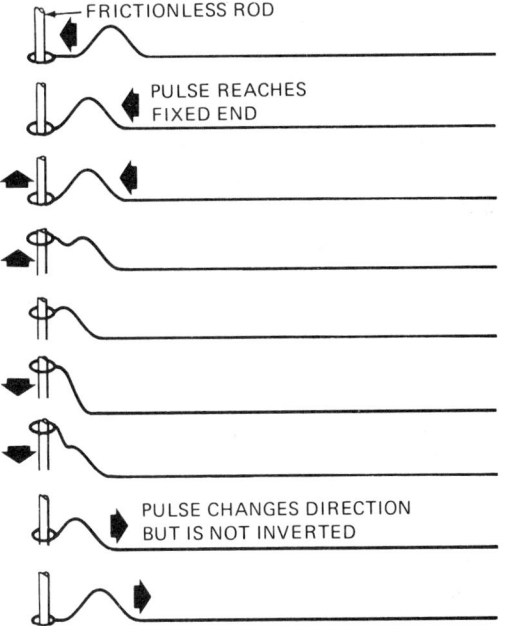

Fig. 21-6

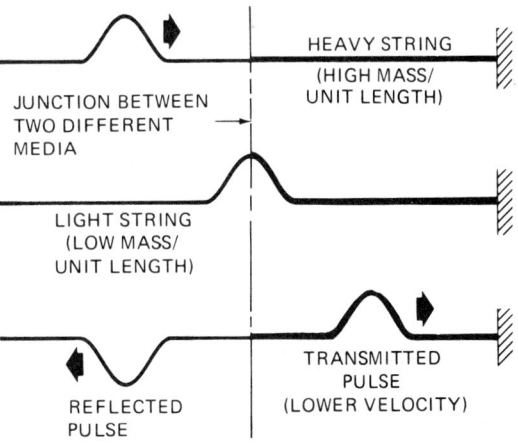

Fig. 21-7

■ PRINCIPLE OF SUPERPOSITION

The previous paragraphs considered a pulse as a single train (a single pulse transmitted from a single source along a medium). When two or more pulses move along a medium toward each other, there is a larger pulse at the instant the pulses come together, figure 21-8. The separate pulses then reappear unchanged and continue in the original directions of motion. The pulses are unaffected by their crossing. The principle explaining this type of behavior is called *superposition*. This principle states that when two or more wave trains (pulses) exist simultaneously in the same medium, the displacement at that point where the pulses meet is equal to the algebraic sum of the instantaneous displacements of each wave. In other words, each wave travels through the medium as if the other wave were not present.

When two identical pulses exist in the same medium, but one pulse is inverted with respect to the other pulse, the displacements cancel out, figure 21-9. However, after complete cancellation, the pulses reappear unchanged and continue in their original directions of motion.

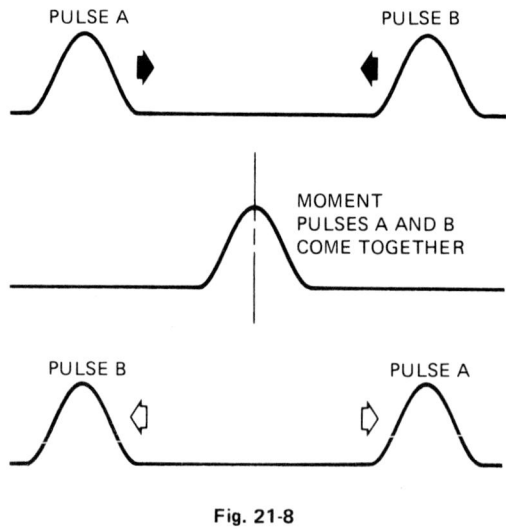

Fig. 21-8

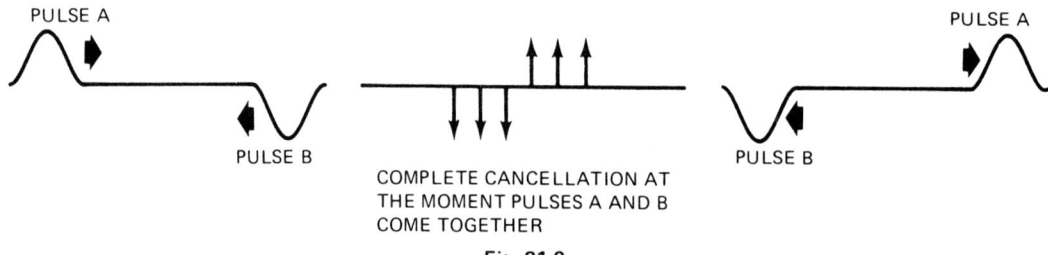

Fig. 21-9

Constructive and Destructive Interference

When two waves of the same wavelength come together (coincide) so that the crests and troughs of the waves meet, the waves are said to *interfere constructively*, figure 21-10A. The amplitude of the resulting wave is greater than the amplitude of either of the original waves. According to the superposition principle, the amplitude is equal to the algebraic sum of the instantaneous amplitudes of the individual waves.

Similarly, when two waves of equal wavelength come together so that a crest meets (coincides with) a trough and the trough meets a crest, the waves *interfere destructively*, figure 21-10B. The amplitude of the resultant composite wave is less than the amplitude of the larger of the two original waves.

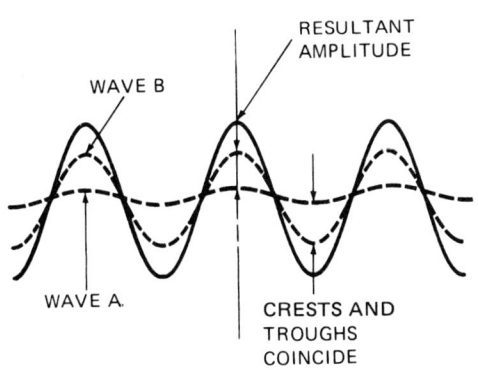
(A) CONSTRUCTIVE INTERFERENCE

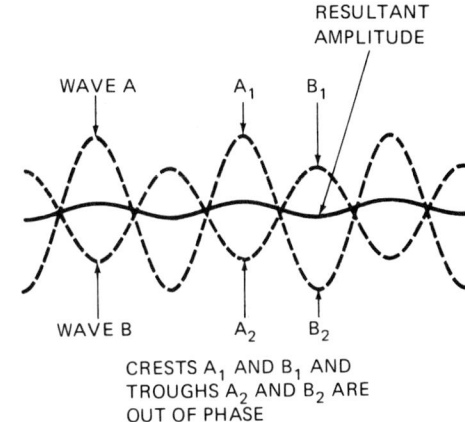
(B) DESTRUCTIVE INTERFERENCE

Fig. 21-10

■ PERIODIC WAVES

A *periodic wave* may describe sound, light, water, and other types of waves where one pulse follows another in regular succession. The shape of the respective waves (the *waveform*) is repeated at regular intervals. A wave train of periodic waves consists of a number of crests and troughs which move at a constant rate of speed.

The distance between two adjacent crests or troughs in the wave train is called the *wavelength*. The symbol for wavelength is the Greek letter lambda (λ). In a periodic wave train the wavelength (λ) is the distance between two particles that are in phase. Two particles are *in phase* when they are moving in the same direction and have the same displacement at the same time.

Description of Periodic Waves

Periodic waves are described by three related quantities: wave velocity (v), wavelength (λ), and frequency (f). The distance a wave moves per second (*wave velocity*) equals the number of waves passing a given point per second (*frequency*) times the distance between adjacent crests or troughs of the wave (*wavelength*).

$$\text{Wave velocity (v)} = \text{frequency (f)} \times \text{wavelength } (\lambda)$$

$$v = f \times \lambda$$

The *period* of a wave is equal to the time (T) required for one wavelength to pass a given point.

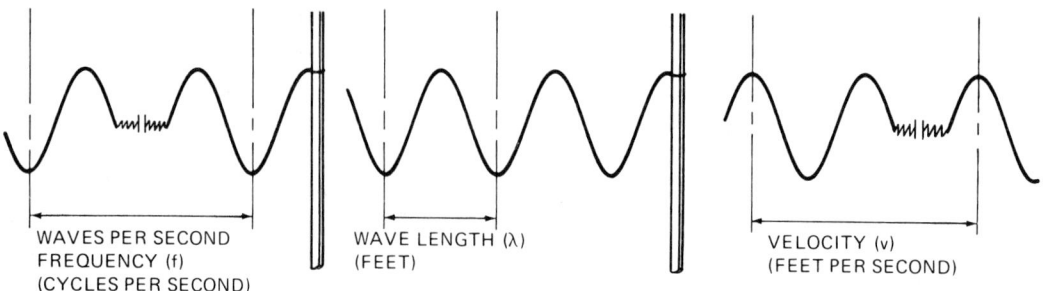

Fig. 21-11

$$\text{Time (seconds)} = \frac{1}{\text{frequency}}$$

$$T = \frac{1}{f}$$

Wave velocity (v) = frequency (f) × wavelength (λ)

$$= \frac{\text{frequency (f)}}{\text{period of wave (T)}}$$

Frequency is measured in units of waves per second or cycles per second. In SI metrics, the hertz (Hz) is the unit of frequency. Wavelength is expressed by a linear unit and velocity in linear units per second, figure 21-11.

Frequency can be expressed as follows:

$$\text{1 kilocycle per second (kc/s)} = \text{1 kilohertz (kHz)} = 10^3 \text{ c/s}$$

$$\text{1 megacycle per second (Mc/s)} = \text{1 megahertz (MHz)} = 10^6 \text{ c/s}$$

For example, 40 MHz = 40 × 10^6 Hz = 4 × 10^7 c/s.

The following example illustrates the procedure for determining the frequency, velocity, and wavelength of periodic waves.

The distance between wave crests is 80 feet. In a one-minute interval, ten waves pass a given point. Determine the frequency and velocity of the waves.

$$f = \frac{1}{T(s)} = \frac{1}{6} = 0.17 \text{ Hz}$$

$$v = \frac{\lambda}{T} = \frac{80}{1/6} = 480 \text{ ft/s or } 146.4 \text{ m/s}$$

Amplitude

A wave can also be described in terms of its amplitude. As shown in figure 21-12, *amplitude* is the height of the wave crest or the depth of the wave trough as measured from the original position of the particle. Thus, the amplitude (A) of a wave is its maximum displacement from the normal position of the particle as it oscillates (moves back and forth) while the wave moves.

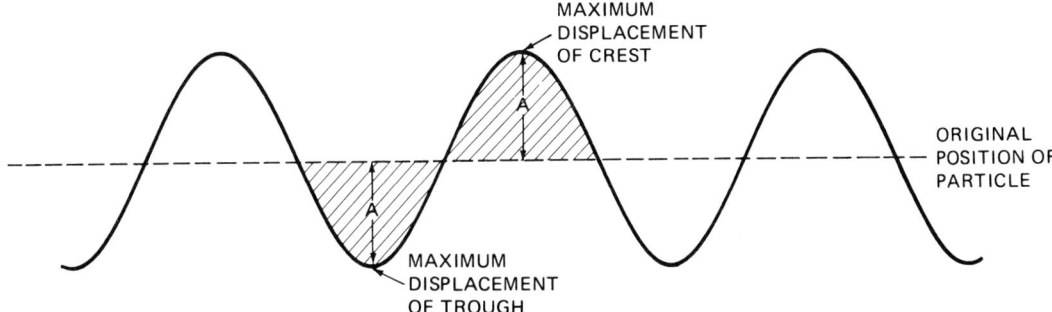

Fig. 21-12 Amplitude (A) of a wave

This unit stresses the importance of understanding the principles of wave motion in their application to the transfer and transmission of other forms of energy.

SUMMARY

- Wave motion is an important mode of transporting energy from one place to another. Wave motion is caused by a disturbance in the medium.
- Transverse and longitudinal waves are the two major types of waves. The vibration of the individual particles in a medium is:
  - perpendicular to the motion of the wave in transverse waves, and
  - for longitudinal waves, parallel to the direction of motion of the waves in the medium.
- Removal of a force causing condensation results in the propagation of a condensation pulse. A rarefaction pulse is propagated by removing the force producing rarefaction.
- The velocity of a pulse $= \sqrt{\dfrac{\text{tension}}{\text{mass/length}}} = \sqrt{\dfrac{T}{m/l}}$.
- When a pulse reaches a fixed end, the pulse is subjected to an equal and opposite reaction force. This force produces an inverted pulse moving in the opposite direction.
- When two or more waves of the same type travel past a given point at the same time, the resultant amplitude is equal to the sum of the amplitude of each wave.
- Interference is the interaction of different wave trains.
  - Constructive interference occurs when the amplitude of the composite wave is greater than the amplitudes of the original waves.
  - Destructive interference occurs when the amplitude of the composite wave is less than the amplitudes of the original waves.

- Periodic waves are described by the related quantities wave velocity (v), wavelength (λ), and frequency (f).

$$\text{velocity} = \text{frequency} \times \text{wavelength}$$
$$v = (f)(\lambda)$$

- The period of a wave is measured by the time (T) in seconds required for one wavelength to pass a given point.

$$\text{Time (s)} = \frac{1}{\text{frequency (f)}}$$
$$T = \frac{1}{f}$$

- Frequency is measured in waves or cycles per second (c/s) or hertz (Hz).

$$\text{Frequency} = \frac{1}{\text{time (s)}}$$
$$f = \frac{1}{T}$$

- Wavelength is expressed in a linear measurement; velocity is expressed in feet per second (ft/s).

$$\text{Velocity} = \frac{\text{wavelength}}{\text{time}}$$
$$v = \frac{\lambda}{T} = (f)(\lambda)$$

- The amplitude of a wave is the maximum height or depth that a particle of the medium is displaced on either side of the normal position of the medium as the wave passes.

ASSIGNMENT UNIT 21 WAVE MOTION: TRANSFER OF ENERGY

■ PRACTICAL PROBLEMS WITH WAVE MOTION AND PHENOMENA

1. Define each of the following terms or quantities:
   a. Mechanical wave
   b. Wave motion
   c. Transverse wave
   d. Longitudinal wave
   e. Condensation pulse
   f. Rarefaction pulse
   g. Constructive interference
   h. Frequency
   i. Amplitude

Of the choices given for statements 2-9, determine which one(s) correctly complete(s) each statement.

2. The individual particles of a transverse wave move in a direction that is _____ _____ to the direction of travel of the wave.
   a. parallel
   b. perpendicular
   c. circular
   d. opposite

3. In longitudinal waves, the individual particles of a medium vibrate back and forth in the _____ direction as that in which the waves travel.
   a. same
   b. reverse
   c. perpendicular
   d. elliptical

4. The instantaneous amplitudes of two waves that meet at a specific time are A and B. The combined amplitude is
   a. $\frac{A + B}{2}$.
   b. A + B.
   c. A - B.
   d. $\frac{A - B}{2}$.

5. The period of a wave is the time required for
   a. one pulse.
   b. a number of waves to pass a point per second.
   c. each wave to move a specific distance per second.
   d. one complete wave to pass a given point.

6. Three related quantities that are used to describe periodic waves are:
   a. distance
   b. frequency
   c. constructive interference
   d. crests
   e. wavelength
   f. wave velocity

7. At the junctions between different mediums, all types of waves exhibit
   a. constructive interference.
   b. destructive interference.
   c. instantaneous amplitude.
   d. reflection and transmission.

8. The maximum displacement from the normal position of the moving particles as a wave passes is identified as the _____ of a wave.
   a. frequency.
   b. wavelength.
   c. velocity.
   d. amplitude.

9. Wave velocity equals
   a. frequency × wavelength.
   b. the sum of the displacement.
   c. frequency × cycles.
   d. frequency × amplitude.

10. Make generalized statements to explain each of these phenomena:
    a. In a uniform metal wire under tension, a pulse reaching a fixed end is inverted upon reflection.
    b. When a pulse reaches a free end of a metal wire, it is not inverted upon reflection.
    c. In two strings of different diameters but under the same tension, a pulse travels slower in the heavier string.

11. State the principle of superposition. Illustrate the principle by a simple line drawing.

12. Compute the speed of longitudinal waves A, B, C, and D. The frequency and wavelength for each wave is given in the table. Round off values C and D to the nearest whole number.

Wave	Frequency (f) Cycles/s	Wavelength	Wave Velocity (v)
A	100	6 ft	ft/s
B	20×10^3	10 ft	ft/s
C	365	9.5 m	m/s
D	29.2	6.8 m	m/s

Achievement Review of Mechanics, Machines, and Wave Motion

■ FORCES AND THEIR EFFECTS

Indicate the letter corresponding to the item which correctly completes statements 1-7.

1. A number of forces acting on a body act
   a. one at a time.
   b. in combination.

2. Effort force (E) × effort distance (ED) =
   a. E × R.
   b. R × ED.
   c. R × RD.
   d. RD × ED.

3. The mechanical advantage of force where an effort force of 20 kg moves a 100-kg load is
   a. 50.
   b. 5.
   c. 1/5.
   d. 2000.

4. The mechanical advantage of speed for a machine whose resistance moves through a distance of 12.4 meters to move an effort force 3.1 meters is
   a. 4.
   b. 1/4.
   c. 32.

5. Work output equals
   a. E × ED.
   b. E × R.
   c. R × RD.

6. Efficiency equals
   a. output × input.
   b. E × ED.
   c. output ÷ input.

7. Horsepower, the measure of the rate of doing work, equals
   a. 550 × work × time.
   b. $\dfrac{\text{Work (Foot-pounds)} \times \text{Time (Min)}}{33\,000}$.
   c. $\dfrac{33\,000}{\text{Work} \times \text{Time}}$.

8. Compute the horsepower needed to do the work indicated in the table at A, B, C, and D.

	Rate of Work	Horsepower
A	99 000 ft-lb/min	
B	1100 ft-lb/s	
C	28 825 ft-lb/min	
D	137.5 ft-lb/s	

■ BALANCE, EQUILIBRIUM – PARALLEL AND ANGULAR FORCES

For statements 1 to 5, determine which are true (T) and which are false (F).

1. A dynamically balanced part is not balanced while in motion.
2. The vectors of a vector diagram indicate both the direction and magnitude of one or more forces.
3. An object is stable when its center of gravity falls within its base.
4. The resultant of two forces acting in directly opposite directions is the sum of the forces.
5. A single force that is equal and opposite to a resultant force is the equilibrant.
6. Select a scale to represent the values of the vectors in figure 1 at A, B, and C.
   a. Lay out the vector diagrams accurately.
   b. Determine the value of the resultant in each case to the nearest whole number.

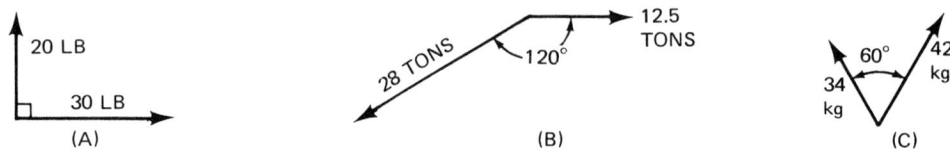

Fig. 1

7. Compute the resultant mathematically for the problems represented in figure 1.
   a. Check the computed values (to the nearest whole number) against the scaled values.
   b. Rework any value that is different.

■ GRAVITATION, MOTION, AND MECHANICAL MOVEMENTS

Complete statements 1 to 3 correctly by indicating the missing word, phrase, or value.

1. Regardless of the weight of a body, a freely falling body gains speed uniformly at the rate of _____.
2. Mass × _____ equals change in momentum.
3. Rectilinear motion may be (a) _____, (b) _____, or any combination of these.
4. Determine the final speed and the average speed for each of the three engines in the table, according to the conditions given.

Engine	Starting Speed (Rev.)	Uniform Acceleration (rps)	Time to Reach Final Speed (s)	Final Speed (Rev.)	Average Speed (Rev.)
A	Rest	100	10		
B	1000	100	10		
C	Rest	900	3.5		

5. Identify two rotating parts, machines, or mechanism where the manufacturer must dynamically balance the moving parts to overcome any centrifugal force.

■ SIMPLE MACHINES: LEVERS

Select the correct word or phrase to complete statements 1, 2, and 3.

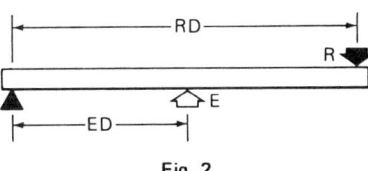

Fig. 2

1. A (an) (wedge) (inclined plane) (lever) is a rigid bar which turns around a fulcrum.
2. Of the three major classes of levers, (Class 1) (Class 2) (Class 3) has the resistance force between the fulcrum at one end and the effort force.
3. Figure 2 represents a (Class 1) (Class 2) (Class 3) lever.
4. Find E, ED, or RD, as required, for levers A, B, and C. Neglect the weight of the lever and prove each answer.

Lever	Class	E	ED	R	RD
A	1	62.5 lb	8 ft	40 lb	
B	2		6.86 m	125 tons	3.66 m
C	3	648 lb		540 lb	9 ft

5. a. For a mechanical or electrical device, name two parts that serve as levers.
   b. Identify each lever as to class by making a simple diagram and labeling the important parts.
   c. Describe briefly the function of each of the two levers.

■ SIMPLE MACHINES: INCLINED PLANE AND WEDGE

For statements 1 to 4, determine which are true (T) and which are false (F).
1. The larger the angle of the inclined plane, the greater is the effort force needed to produce work.
2. Tapered holding devices are held securely by a wedging action.
3. The effort force on a wedge moves parallel to the slope.
4. The mechanical advantage of force of an inclined plane is equal to RD/ED.
5. Compute the forces or distances missing in the table for inclined planes A, B, and C. Neglect friction. Round off answers to the nearest whole number.

	Effort Force (E)	Slope Length (ED)	Load (R)	Height (RD)
A		14 m	6 customary tons	4 m
B	145 kg		577 kg	1.5 m
C		4.7 m	4.75 metric tons	1.4 m

6. Compute the resistance force for the four wedge angles given in the table.

Wedge Angle	Drive Force (E)	Effort Movement (ED)	Resistance Force (R)	Follower Distance (RD)
20°	200 lb	1 in		1/2 in
40°	200 lb	1 in		1 in
60°	200 lb	1 in		1 1/2 in
80°	200 lb	1 in		2 in

7. Plot a graph of the resistance force (R) of a wedge-shaped cam from the information in the table given in question 6. Use a vertical scale of 1/4" = 5°, and a horizontal scale of 1/4" = 20 pounds.

8. Refer to the resistance force and wedge angle graph plotted in question 7.
   a. Determine the value of the resistance force for wedge angles of 30°, 45°, and 75°.
   b. Explain the effect on the resistance force of increasing the wedge angle from 20° to 80°.

■ SIMPLE MACHINES: THE WHEEL AND AXLE

1. The following table gives distances and forces for four wheel and axle combinations.
   a. Make a wheel and axle diagram, labeling (R), (E), (ED), and (RD).
   b. Compute the missing values.

	Wheel and Axle Combinations			
	E	ED	R	RD
A	25 kg	2.4 in	60 kg	
B	158 kg	30.48 cm		39.37 cm
C		0.6 m	2.6 tons	0.4 m
D	1584 kg		1900 kg	47 cm

2. a. Determine the mechanical advantage of speed of the four wheel and axle combinations given in problem 1.
   b. Determine the mechanical advantage of force.

■ SIMPLE MACHINES: THE SCREW THREAD

1. Screw threads are used: (a) to increase force, (b) as measuring devices, (c) to change the direction of motion, and (d) as fasteners. Give two new applications of each of the four uses of screw threads.

2. Compute the load which may be raised by the construction jacks A, B, C, and D for the dimensions and forces shown in the table. Disregard friction losses.

Jack	Threads Per In	Lever Radius (ED)	Effort Force (E)	Load (R)
A	4	35 in	80 lb	
B	3	35 in	80 lb	
C	2	35 in	80 lb	
D	10	35 in	80 lb	

3. Prepare a graph showing the relationship between threads per inch and load (R) as given in the table of problem 2.
   a. Use a horizontal scale of $1'' = 2$ threads.
   b. Use a vertical scale of $1'' = 50\ 000$ lb.

4. Determine from the graph in problem 3 what loads may be raised by changing the threads per inch to 6, 8, and 12.

■ SIMPLE AND COMPOUND GEAR TRAINS

1. Determine the direction of rotation of each of the simple and compound gear trains shown in the table.

	Gear Train (Dr = Driver) Available Gears: Two Each of 24, 30, 32, 36, 48, 56, 64, and 96 Teeth	Speed Ratio Driver to Final Driven Gear
A		1 to 1
B		4 to 1
C		4 to 1
D		6 to 1

2. Select the gears from the given sizes which will produce the required speed ratios for trains A, B, C, and D.

3. Compute the speed in rpm of the final driven gear using initial driver speeds of 40 rpm, 200 rpm, 320 rpm, and 400 rpm respectively.

■ SIMPLE AND COMPOUND MACHINES

Match the numbered items on the left with the lettered items on the right.

1. MA_f of a compound machine
2. MA_f of pulleys
3. A pulley
4. The worm and worm wheel

a. A compound machine consisting of the wheel and axle and the inclined plane.
b. The product of the MA_f of each separate machine.
c. The number of grooves in a pulley.
d. The number of ropes supporting the load on the movable pulley.
e. A change in the direction and magnitude of a force.
f. An application of the lever in a circular shape.

5. A single pitch worm gear revolving at 1440 rpm requires an effort force of 12 lb to drive a 96 T gear, figure 3. Mounted on the same solid shaft is a 32 T gear which drives a 128 T gear.
   a. Determine the resistance which this gear combination can move.
   b. Compute the mechanical advantage of force.
   c. Compute the mechanical advantage of speed.

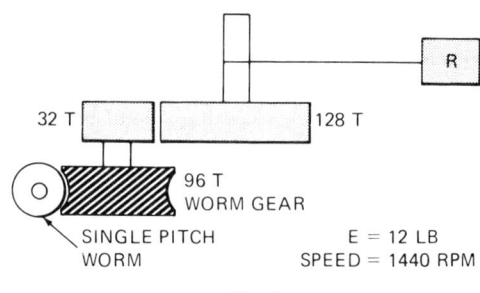

Fig. 3

■ MECHANICAL POWER TRANSMISSION, FRICTION, AND LUBRICATION

Indicate the correct word(s) or phrase to complete statements 1, 2, and 3.

1. When the source of power is brought close to the driven unit, some _____ _____ losses are eliminated.

2. Friction is caused by _____, _____, _____, _____, or _____.

3. Friction may be reduced by _____, _____, _____, _____, or _____.

4. Select a common machine and name three parts which depend upon friction for their operation.

5. The results of a Prony brake test for three motors are shown in the table.
   a. Compute the amount of work done by each motor in one minute.
   b. Convert the work value to the nearest fractional horsepower for each motor.
   c. Compute the equivalent horsepower in SI metric joules per second units for each motor. Use 1 ft-lb = 1.4 joules.

Motor	Force (Lb)		Diam.	Rpm	Work (1 min)	Hp	Joules (J)*
	F_1	F_2					
A	12	6	7 in	1000			
B	12	6	3 1/2 in	2000			
C	26.4	12.6	2.5 in	5600			

   *1 ft-lb = 1.4 J

■ MECHANICS OF FLUIDS AT REST AND IN MOTION

Select the word or words which correctly complete statements 1 to 4.
1. Gravity pressure is (an externally applied pressure) (an internal pressure exerted by the weight of the fluid).
2. Hydraulic machines require the combined application of (force, distance, and pressure) (force and pressure) (pressure and distance).
3. The pressure of a fluid (does not change) (increases) (decreases) whenever the speed of a liquid is increased as it passes a constriction.
4. The (specific gravity) (gravity pressure) (force) of sinking and floating bodies equals the weight of the body divided by the weight of an equal volume of water.
5. Compute the force on the bottom of tanks, A, B, C, and D. Use the weight of water as 62.4 lb/ft^3 or 1005 kg/m^3, as required.

	Volume of Fluid			Specific Gravity	F	P
	L	W	D			
A	1 m	1 m	1 m	0.5	kg	kg/m^2
B	1 m	1 m	10 m	0.5	kg	kg/m^2
C	Diameter 3 1/2 ft		5 ft	1.0	lb	lb/ft^2
D	Radius 14 ft		9 in	0.8	lb	lb/ft^2

6. Determine the pressure on tanks, A, B, C, and D, using the values from the table.

7. Refer to the drawing of the lever-actuated hydraulic lift. Use the given dimensions and the data in the table to find the missing values.

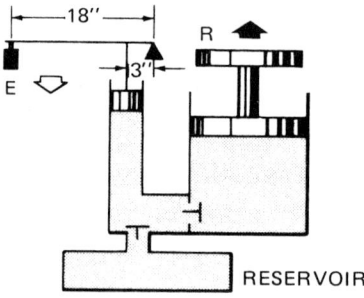

Fig. 4

	Resistance Force (R)	Diameter Large Piston	Diameter Small Piston	Effort Force
A		4 in	2 in	20 lb
B	1200 lb	10 in	2 in	
C	12 000 lb		1 in	100 lb
D	120 tons	20 in	1 in	

ATMOSPHERIC PRESSURE: PRINCIPLES AND APPLICATIONS

1. Determine the heights to which the columns of Fluids I, II, and III can be raised for the pressures given at A, B, and C. As the standard, use the weight of water (62.4 lb/cu ft) and a force of 1.033 2 kg/cm² at atmospheric pressure to support a column of water 10.2 meters high. Round off answers to A and B to one decimal place; round off C to two decimal places.

	Pressure	Fluid I	Fluid II	Fluid III
			Specific Gravity	
		Water	0.6	2.5
A	14.7 lb/sq in	ft	ft	ft
B	100 lb/sq ft	ft	ft or in	ft or in
C	8.94 kg/cm²	m	m	m

2. Compute the volume of gases A, B, C, and D when the pressures are reduced as indicated in the table. Round off to the nearest whole number.

	Pressure (P)	Volume 1	Volumn 2 When P is Reduced to	Hourly Consumption	Gas Supply
	Lb/Sq In	Cu In	15 Lb/Sq In	Cu In	Hours
A	30	1000		250	
B	60	1000		500	
	Lb/Sq Ft	Cu Ft	2160 Lb/Sq Ft	Cu Ft	Hours
C	4320	1000		40	
	Tons/Sq Ft				
D	12.6	1000		33	

3. Use the computed volume (V) and the hourly consumption of each gas and determine the length of time each gas can be used. Round off each answer to the nearest hour.

■ FLUID POWER: PRINCIPLES AND APPLICATIONS

1. Describe the differences among hydraulic, pneumatic, and fluid power systems.
2. Identify three design factors, conditions, or phenomena which affect fluid flow.
3. Illustrate a simple series fluid power circuit and a simple parallel circuit. Name the major components.
4. Name and describe two types of motion that are characteristic of fluid flow.
5. A parallel fluid system has four branches. The resistances in each branch are given in the table. Determine the total resistance of the system.

Branch	Resistance (N/m^2-min/L)
A	16
B	24
C	32
D	48

6. Three separate series circuits have the pressures and flow rates shown in the table. Determine the resistance of each circuit.

Circuit	Pressure N/m²	Flow Rate (Gal/Min)	Resistance (N/m²–min/L)
A	396	8.4	
B	438	9.6	
C	724	10.2	

7. Use the resistance values given in the table for the components of a series fluid circuit having a flow of eight gallons per minute.
   a. Compute the total resistance for each set of components.
   b. Find the total resistance for all components.
   c. Determine the pressure drop for each set of components.
   d. Determine the pressure drop for the whole system.

	Component	Resistance Unit Value	Resistance Total for All Components	Pressure Drop (Psi-min/Gal)
A	60 ft of hydraulic hose, 1/4-in diam.	1.8		
B	Pressure gages (3)	5.4		
C	Flowmeters (2)	0.9		
D	Quick disconnect couplers (8)	5.2		
	Total for System			

8. Determine the (a) potential energy and (b) the kinetic energy of 608 cubic feet of water in a storage tank. The fluid passes through an outlet at the rate of six feet per second at the beginning of the process. At the end of the process, the water passes through the outlet at the rate of four feet per second.

H_1 = 40 in

H_2 = 20 in

g = 32 ft/s²

$M_1 = \dfrac{4.23 \text{ lb-s}^2}{\text{ft}}$

$M_2 = \dfrac{2.16 \text{ lb-s}^2}{\text{ft}}$

■ WAVE MOTION: TRANSFER OF ENERGY

1. Draw two waves having the same wavelength and different amplitudes. The crests and troughs of the waves are to coincide. Describe the magnitude of the composite wave and the conditions that are created.

2. Calculate the equivalent values in cycles/second for A, B, C, and D, according to the values given in the table.

Medium	Frequency (f)	Equivalent Value in Cycles/second
A	60 megahertz (MHz)	
B	30 kilocycles (kc/s)	
C	25 megacycles (Mc/s)	
D	2.5 kilohertz (kHz)	

3. The distance between adjacent wave crests (λ) and the time interval of waves A, B, C, and D are given in the table. Determine both the frequency and the velocity of each wave.

Wave	Wavelength (λ)	Time (s) per Passing Event	Frequency (f)	Velocity (v)
A	80 ft	2.25		
B	160 ft	2.25		
C	80 ft	4.5		
D	160 ft	4.5		

Appendix

TABLE I STANDARD TABLES OF METRIC UNITS OF MEASURE

Linear Measure

Unit	Value in Meters	Symbol or Abbreviation
micron	0.000 001	μ
millimeter	0.001	mm
centimeter	0.01	cm
decimeter	0.1	dm
meter (unit)	1.0	m
dekameter	10.0	dam
hectometer	100.0	hm
kilometer	1 000.00	km
myriameter	10 000.00	mym
megameter	1 000 000.00	Mm

Surface Measure

Unit	Value in Square Meters	Symbol or Abbreviation
square millimeter	0.000 001	mm^2
square centimeter	0.000 1	cm^2
square decimeter	0.01	dm^2
square meter (centiare)	1.0	m^2
square dekameter (are)	100.0	dam^2
hectare	10 000.0	ha^2
square kilometer	1 000 000.0	km^2

Volume

Unit	Value in Liters	Symbol or Abbreviation
milliliter	0.001	mL
centiliter	0.01	cL
deciliter	0.1	dL
liter (unit)	1.0	L
dekaliter	10.0	daL
hectoliter	100.0	hL
kiloliter	1 000.0	kL

Mass

Unit	Value of Grams	Symbol or Abbreviation
microgram	0.000 001	μg
milligram	0.001	mg
centigram	0.01	cg
decigram	0.1	dg
gram (unit)	1.0	g
dekagram	10.0	dag
hectogram	100.0	hg
kilogram	1 000.0	kg
myriagram	10 000.0	myg
quintal	100 000.0	q
ton	1 000 000.0	t

Cubic Measure

Unit	Value in Cubic Meters	Symbol or Abbreviation
cubic micron	10^{-18}	μ^3
cubic millimeter	10^{-9}	mm^3
cubic centimeter	10^{-6}	cm^3
cubic decimeter	10^{-3}	dm^3
cubic meter	1	m^3
cubic dekameter	10^3	dam^3
cubic hectometer	10^6	hm^3
cubic kilometer	10^9	km^3

TABLE II CONVERSION FACTORS USING CUSTOMARY AND SI METRIC PHYSICAL CONSTANTS*

Length	1 meter (m) = 39.37 in, 3.281 ft, 1.094 yd 1 centimeter (cm) = 0.393 7 in 1 millimeter (mm) = 0.039 37 in 1 kilometer = 0.621 4 mile (mi)	1 inch (in) = 0.083 3 ft, 2.54 cm, 25.4 mm 1 foot (ft) = 0.304 8 meter (m) 1 mile (mi) = 1.609 kilometers (km)
Area	1 cm^2 = 10^{-4} m^2, 0.155 in^2 1 m^2 = 10^4 cm^2, 1.55 × 10^3 in^2, 10.76 ft^2, 1.196 yd^2 1 km^2 = 0.386 1 mi^2 1 hectare (ha) = 10 000 m^2, 2.471 acres	1 in^2 = 6.452 cm^2 1 ft^2 = 9.29 × 10^{-2} m^2 (0.092 9 m^2) 1 acre = 0.404 7 ha 1 mi^2 = 2.59 km^2
Volume	1 cm^3 = 0.061 in^3, 0.001 0 liter (L) 1 L = 61 in^3, 1000 cm^3, 1.06 qt, 0.264 gal 1 m^3 = 10^6 cm^3, 10^3 liters (L), 35.31 ft^3, 1.308 yd^3, 6.10 × 10^4 in^3	1 liquid ounce (oz) = 29.57 cm^3 1 quart (qt) = 57.8 in^3, 0.946 L 1 U.S. gal = 0.134 ft^3, 3.78 L, 3.785 × 10^{-3} m^3 1 in^3 = 16.39 cm^3, 0.17 3 qt 1 ft^3 = 2.83 × 10^{-2} m^3 (0.028 3 m^3), 28.3 L, 7.481 gal 1 yd^3 = 0.764 6 m^3
Mass	1 kilogram (kg) = 0.068 5 slug, 35.3 oz, 2.2 lb (weight/kilogram), 10^3 gram 1 slug = 32.17 lb, 14.59 kg 1 ton (t) = 907 kg	1 oz = 28.4 g, 194 × 10^{-3} slug 1 lb mass = 454 g, 0.454 kg, 3.11 × 10^{-2} slug
Velocity	1 m/s = 3.28 ft/s, 2.24 mi/h, 3.60 km/h 1 km/h = 0.278 m/s, 0.913 ft/s, 0.621 4 mi/h	1 ft/s = 0.305 m/s, 1.10 km/h, 0.682 mi/h 1 mi/h = 1.467 ft/s, 0.447 m/s, 1.61 km/h
Force	1 newton (N) = 0.225 lb, 3.60 oz, 10^5 dynes	1 lb = 4.45 N, 4.45 × 10^5 dynes
Pressure	1 N/m^2 = 1.45 × 10^{-4} lb/in^2, 2.09 × 10^{-2} lb/ft^2 1 atmosphere (atm) = 1.013 × 10^5 N/m^2, 14.7 lb/in^2 1 kPa = 6.9 lb/sq in	1 lb/in^2 = 144 lb/ft^2, 6.90 × 10^3 N/m^2 14.7 lb/in^2 = 1.033 kg$_f$/cm^2 = 101.33 kPa
Energy	1 joule (J) = 0.738 ft-lb, 2.39 × 10^{-4} kcal, 6.24 × 10^{18} eV, 3.73 × 10^{-7} hp-h 1 kilocalorie (kcal) = 4185 J, 3.97 Btu, 3077 ft-lb	1 ft-lb = 1.36 J, 1.29 × 10^{-3} Btu, 3.25 × 10^{-4} kcal 1 Btu = 778 ft-lb, 0.252 kcal, 2.93 × 10^{-4} kWh 1 Btu/lb = 0.555 kcal/kg
Power	1 watt (W) = 1 J/s, 0.738 ft-lb/s 1 kilowatt (kW) = 1.341 hp	1 horsepower (hp) = 746 W, 550 ft-lb/s, 2550 Btu 1 refrigeration ton = 12 000 Btu/h
Temperature	$T_K = T_C + 273°$ $T_R = T_F + 460°$	$T_C = 5/9\ T_F - 32°$ $T_F = 9/5\ T_C + 32°$
Angle	1 radian (rad) = 57° 18', 57.30° 1 rad/s = 9.55 rpm	1° = 0.017 45 rad 1 rpm = 0.104 7 rad/s

*Important Note: Converted measurements among customary inch and SI metric units are based on either *exact table values* or rounded-off values. For practical purposes, *exact values* are used for precise measurements; rounded-off values are used for general explanations, examples, and applications.

TABLE III SYMBOLS AND SELECTED DERIVED UNITS OF PHYSICAL QUANTITIES

Quantity	Symbol	Derived Units*
Acceleration	a	meters/second^2
Angular acceleration	α	radians/second^2
Angular displacement	θ	radian
Angular momentum	L	kilogram-meters^2/second
Angular velocity	ω	radians/second
Area	A	meter^2
Density, mass	ρ	kilograms/meter^3
Energy, heat	kcal	kilocaloric
total	J	joule
Force	N	newton
Gravitational field intensity	γ	newtons/kilogram
Linear momentum	p	kilogram-meters/second
Pressure	Pa	newtons/meter^2
Rotational inertia	I	kilogram-meter^2
Torque	τ	newton-meter
Velocity, speed	v	meters/second
Volume	V	meter^3
Work	J	joule

*The derived units are stated in the SI metric system.

TABLE IV UNITS OF MASS AND WEIGHT AND CONVERSION FACTORS

System	Units - Mass	Units - Weight	Acceleration of Gravity (g)	Conversion Factors - Mass (m) (Given: Weight)	Conversion Factors - Weight (w) (Given: Mass)
Metric	kilogram (kg); 1 kg = 0.068 5 slug	Newton (N); 1 N = 0.225 lb	9.8 m/s^2	$m \text{ (kg)} = \dfrac{w(N)}{9.8 \text{ m/s}^2}$	$w \text{ (N)} = m \text{ (kg)} \times 9.8 \text{ m/s}^2$
British	slug; 1 slug = 14.6 kg	pound (lb); 1 lb = 4.45 N	32 ft/s^2	$m \text{ (slugs)} = \dfrac{w(\text{lb})}{32 \text{ ft/s}^2}$	$w \text{ (lb)} = m \text{ (slugs)} \times 32 \text{ ft/s}^2$

TABLE V COMPARATIVE DENSITIES OF GASES USED IN INDUSTRY

	Gas	Density (lb/ft^3)	Specific Weight
Lighter than air	Hydrogen	0.005	0.07
	Methane (natural gas)	0.045	0.55
	Ammonia	0.045	0.60
	Acetylene	0.068	0.91
	Carbon monoxide	0.073	0.98
	Nitrogen	0.074	0.98
	Air (approx.)	0.075	1.00
Heavier than air	Formaldehyde	0.077	1.03
	Oxygen	0.084	1.12
	Ethane	0.085	1.05
	Hydrogen sulphide	0.096	1.28
	Carbon dioxide	0.116	1.61
	Propane (bottled cooking gas)	0.117	1.56
	Nitrogen peroxide	0.128	1.71
	Isobutane	0.151	2.01
	Butane	0.155	2.05
	Sulphur dioxide (refrigerant)	0.183	2.44
	Chlorine	0.183	2.44
	Gasoline vapors (octane)	0.290	3.86

TABLE VI MASS AND WEIGHT DENSITIES OF COMMON SUBSTANCES*

Substance	Weight Density lb/ft^3	Mass Density kg/m^3	Mass Density g/cm^3	Mass Density slugs/ft^3
Air	8×10^{-2}	1.3	1.3×10^{-3}	2.5×10^{-3}
Alcohol (ethyl)	48	7.9×10^2	0.79	1.5
Aluminum	1.7×10^2	2.7×10^3	2.7	5.3
Carbon dioxide	0.12	2.0	2.0×10^{-3}	3.8×10^{-3}
Concrete	1.4×10^2	2.3×10^3	2.3	4.5
Gasoline	42	6.8×10^2	0.68	1.3
Gold	1.2×10^3	1.9×10^4	19	38
Hardwood (oak)	45	7.2×10^2	0.72	1.4
Helium	1.1×10^{-2}	0.18	1.8×10^{-4}	3.5×10^{-4}
Hydrogen	5.4×10^{-2}	0.09	9×10^{-5}	1.7×10^{-3}
Ice	58	9.2×10^2	0.92	1.8
Iron	4.8×10^2	7.8×10^3	7.8	15
Lead	7×10^2	1.1×10^4	11	22
Mercury	8.3×10^2	1.4×10^4	14	26
Nickel	5.5×10^2	8.9×10^3	8.9	17
Nitrogen	7.7×10^{-2}	1.3	1.3×10^{-3}	2.4×10^{-3}
Oxygen	9×10^{-2}	1.4	1.4×10^{-3}	2.8×10^{-3}
Softwood (balsa)	8	1.3×10^2	0.13	0.25
Water, pure	62	1.00×10^3	1.00	1.94
Water, sea	64	1.03×10^3	1.03	2.00

*At atmospheric pressure and room temperature.

TABLE VII COEFFICIENTS OF FRICTION FOR DIFFERENT SOLIDS

Combinations of Materials	Degree of Lubrication	Coefficient of Friction*
Bronze on bronze	None	0.20
Bronze on cast iron	None	0.21
Bronze on cast iron	Slight	0.16
Cast iron on cast iron	Slight	0.15
Cast iron on wrought iron	None	0.18
Wrought iron on wrought iron	None	0.44
Cast iron on hardwood	None	0.49
Cast iron on hardwood	Slight	0.19
Wrought iron on hardwood	Complete	0.08
Leather on hardwood	None	0.33
Leather on cast iron	None	0.56
Smooth surfaces	Complete	0.04
Metal to metal (rolling)	—	0.002

*The coefficients given are for pressures of 14 to 20 pounds per square inch or 992 g/cm^2 to 1407 g/cm^2 (0.99 kg/cm^2 to 1.41 kg/cm^2). The coefficient may vary from sample to sample.

TABLE VIII HEAT VALUES OF GASES, LIQUIDS, AND SOLIDS

	Solids, Liquids, and Gases	Coefficient of Linear Expansion (1°C)	Specific Heat Capacity*	Heat of Fusion (kcal/kg)	Heat of Vaporization (kcal/kg)	Melting Point (Degrees C)	Boiling Point (Degrees C)
Gases	Air		0.24				
	Ammonia		0.52	84.	327.	−78	−33
	Carbon dioxide		0.22	45.	85.		
	Hydrogen		3.40	14.	108.	−259	−253
	Nitrogen		0.25	6.1	48.	−210	−196
	Oxygen		0.22	3.3	51.	−218	−183
	Steam		0.48				
Liquids	Ethyl alcohol		0.58	25.	204.	−130	78
	Mercury		0.03	2.8	65.	−39	357
	Water		1.00	80.	540.	0	100
Solids	Aluminum	0.000 024	0.21	77.		660	1800
	Brass	0.000 019	0.09			1050	
	Copper	0.000 017	0.09	42.		1083	2300
	Glass (reg.)	0.000 009	0.20				
	Glass (pyrex)	0.000 003					
	Ice	0.000 05	0.51	80.		0	
	Iron	0.000 011	0.11	5.5		1535	3000
	Lead	0.000 029	0.03	5.9		327	1620
	Platinum	0.000 009	0.03	27.		1774	4300
	Quartz	0.000 000 5	0.19				
	Silver	0.000 019	0.06	21.		960	1950
	Sulphur	0.000 064	0.17	13.2		113	445
	Zinc	0.000 026	0.09	28.		419	907

*kcal/kg°C and Btu/lb°F

TABLE IX PHYSICAL PROPERTIES OF IMPORTANT PURE METALS

Element	Chemical Symbol	Density (lb per cu in)	Melting Point (°F)	Coefficient of Linear Expansion (Millionths of an inch per °F)	Electrical Resistivity (Millionths of ohms per cm^3)
Aluminum	Al	0.097 51	1220°	13.3	2.655
Chromium	Cr	0.260	3430°	3.4	13.0
Cobalt	Co	0.32	2720°	6.8	6.24
Copper	Cu	0.324	1980°	9.2	1.673
Gold	Au	0.698	1945°	7.9	2.19
Iron	Fe	0.284	2800°	6.5	9.71
Lead	Pb	0.409 7	621°	16.3	20.65
Magnesium	Mg	0.062 8	1200°	14.0	4.46
Mercury	Hg	0.489 6	−38°		94.1
Nickel	Ni	0.322	2650°	7.4	6.84
Platinum	Pt	0.775 0	3225°	4.9	9.28
Silver	Ag	0.379 0	1761°	10.9	1.59
Tin	Sn	0.263 7	450°	13.0	11.5
Tungsten	W	0.697	6200°	2.4	5.5
Vanadium	V	0.217	3200°	4.3	26.0
Zinc	Z	0.258	787°	22.1	5.916

TABLE X TEMPERING AND HEAT COLORS

	Color	Degrees	
		Fahrenheit	Celsius
Temper Colors	Faint straw	400	205
	Straw	440	225
	Deep straw	475	245
	Bronze	520	270
	Peacock blue	540	280
	Full blue	590	310
	Light blue	640	340
Heat Colors	Faint red	930	500
	Blood red	1075	580
	Dark cherry	1175	635
	Medium cherry	1275	690
	Cherry	1375	745
	Bright cherry	1450	790
	Salmon	1550	840
	Dark orange	1680	890
	Orange	1725	940
	Lemon	1830	1000
	Light yellow	1975	1080
	White	2200	1200

Glossary

Absolute Temperature. A temperature reading on the Kelvin (Absolute) scale. The degree absolute represents the same temperature change as the degree Celsius. The Kelvin temperature (K) is related to the Celsius temperature (C) by the formula: °K = °C + 273.

Absolute Zero. A temperature of 0°K, equivalent to −273°C, believed to be the lowest possible temperature that can be obtained.

Acceleration. The rate at which the velocity of a body changes with time. Changes in velocity may be in direction, magnitude, or both.

Acceleration of Gravity (g). The acceleration of a freely falling body near the earth's surface.
$g = 9.8 \text{ m/s}^2$ (metric system)
$g = 32 \text{ ft/s}^2$ (British system)

Adhesion. The attractive force between molecules of different materials.

Amplitude. The distance from a normal position of an object to the final position, as the object vibrates.

Aneroid Barometer. A barometer actuated by an expanding and contracting mechanism instead of by a liquid.

Angle of Incidence. The angle formed by an incident light ray and a perpendicular to the surface.

Antenna. An aerial which receives radio signals and may have various forms of modulation signals.

Archimedes' Principle. The force buoying up an object in a fluid (liquid or gas), equal to the weight of fluid displaced.

Atmospheric Refraction. The effect of the atmosphere in bending light rays.

Atom. The smallest particle of an element containing all of the properties of that element.

Atomic Mass Unit (U). A measurement unit which expresses atomic and nuclear masses.
$U = 1.66 \times 10^{-27}$ kg

Atomic Number. A fixed (constant) number assigned to each element to indicate the number of protons in the nucleus of the atom.

Atomic Weight (See Atomic Mass Unit). A relationship between the weight or mass of an atom of one element and that of an oxygen atom (fixed at 16).

Avoirdupois. A series of weight units based on 16 grams to the ounce and 16 ounces to the pound.

Barometer. An instrument or device for measuring atmospheric pressure.

Base Units (SI). Seven base units are included in SI metrics. These units are: length (meter), mass (kilogram), time (second), electric current (ampere), temperature (kelvin), luminous intensity (candela), and the amount of substance of a system (mole).

Bernoulli's Principle. Any change in the velocity of a fluid caused by a constriction produces an opposite change in the pressure.

Binding Energy of a Nucleus. The energy that is equal to the difference between the mass of a nucleus and the sum of the masses of its individual constituent nucleons.

Boiling Point. The temperature at which a liquid boils under normal atmospheric pressure (76 centimeters of mercury).

Boyle's Law. The volume of a gas at a constant temperature is inversely proportional to its pressure.

British Thermal Unit (Btu). A unit of measurement of heat energy in the British system. A Btu is the amount of heat required to raise the temperature of one pound of water one Fahrenheit degree.

Calorie. A unit of measurement of heat energy in the metric system which equals the amount of heat required to raise the temperature of one gram of water one Celsius degree.

Candela (cd). The basic unit of luminous intensity in the SI metric system and the United States (British) system.

Candlepower. A unit of measurement of illumination equal to the light energy of one standard candle.

Capillary Action. The effect of molecular forces on liquids which cause some liquids to rise in tubes and other liquids to be depressed.

Celsius Degree (°C). Formerly called degrees centigrade in the metric system. Celsius has been adopted as the basic unit of temperature in the SI metric system. A measurement of temperature equal to 1/100 of the distance between the boiling point and the freezing point of water at standard pressure.

Celsius Scale. A system of temperature measurement in which 0° represents the freezing point of water and 100° the boiling point. Each 1/100th part of the scale is equal to one degree Celsius (1°C).

Center of Gravity. The place where the resultant of the gravitational forces on the body acts.

Centrifugal Force. The force which tends to move a rotating body from a circular motion to a straight-line motion.

Centripetal Force. The force which tends to hold a rotating body together, as opposed to centrifugal force.

Charles' Law. The volume of a sample of gas, at constant pressure, is directly proportional to its absolute temperature. V/T = constant at that pressure regardless of changes in V or T, respectively.

Coefficient of Friction. A numerical value equal to the force required to overcome friction divided by the weight of the moving body acting upon a horizontal surface.

Coefficient of Linear Expansion. A numerical value for each different material which indicates its expansion for each degree of temperature change.

Coefficient of Volume Expansion. A numerical value indicating the increase in volume for each degree of temperature change.

Cohesion. The attractive force between like molecules.

Complementary Colors. Those colors of the spectrum which produce white when combined.

Components (Force). Those forces (two or more) which replace a single force as they act together.

Conduction of Heat. The transfer of heat energy from one molecule to the next or from one body to another.

Conductivity. The ability of a material to conduct heat energy (thermal conductivity) or electrical energy (electrical conductivity).

Convection. The transfer of heat by the movement of matter.

Critical Angle. The largest angle of incidence of a light ray in a medium which causes the ray to be refracted and not reflected.

Customary Units. Units defined by the National Bureau of Standards based upon the yard and pound as commonly used in the United States.

Deci (d). An SI metrics prefix signifying a value of one-tenth.

Deka (da). An SI metrics prefix signifying a value of ten times.

Density. The mass of matter per unit of volume.

Derived Units (SI). A series of measurement units in SI metrics which are derived from the seven base units. Derived units extend the use of the base units to meet specific conditions and practical applications in all industrial sectors.

Deuterons. The nuclei of heavy hydrogen atoms (with an atomic weight of two) used in atomic fusion.

Diffraction. The ability of waves to bend around obstacles in their path.

Displacement. The ability of a floating object to displace its own weight.

Ductility. That property of a material which permits it to be drawn into finer and different shapes.

Efficiency. A mathematical ratio of useful work output to total work input.

Elasticity. A property of matter that allows it to return to an original form after being moved out of shape (deformed).

Elastic Limit. The maximum deformation a solid under stress can withstand without being permanently changed.

Energy (E). A quantitative property of matter indicating its capacity to perform work or change some aspect of the physical world. The joule (J) is the metric unit of energy; foot-pounds (ft-lb) is the British unit. Energy is broadly categorized as kinetic, potential, and rest energy.

Equilibrant. That force in a vector diagram which is equal and opposite to the resultant (balances two or more other forces).

Equilibrium. A condition of balance or rest in which the resultant of the forces acting on a body is zero.

Evaporation. A process of transforming a liquid into a vapor.

Fahrenheit Scale. A widely used temperature scale on which the freezing point of water is recorded as 32° and the boiling point is 212° (there are 180 divisions, each one representing one degree, between the two points).

Field, Magnetic. The area through which magnetic forces act.

Fluid. Matter which is in a liquid or gaseous state.

Foot-pound (ft-lb). The unit of work and energy in the British system.

Force (F). Any influence that can cause a body to be moved. The newton (N) is the unit of force in the metric system; the foot-pound (ft-lb) is the unit of force in the British system.

Forced Vibrations. Those vibrations produced in an object when it is forced to vibrate at a frequency other than its natural frequency.

Friction. The resistive forces opposing the motion of two bodies in contact with each other. *Static friction* deals with frictional resistance that must be overcome by a stationary body to be set in motion. *Sliding friction* is the frictional resistance of a body in motion. *Rolling friction* is the friction experienced by a circular body rolling over a smooth flat surface.

Fusion. The process of changing from a solid to a liquid state. Also, *atomic fusion*, in which lighter atoms are combined to produce heavier atoms with a release of energy.

Giga (G). An SI metrics prefix denoting a value of one billion times (10^9).

Gram (g). A derived SI metrics unit of mass and weight. One gram equals 1/1000 kilogram of water at its maximum density.

Gravitation. Every body in the universe attracts every other body with a force that is directly proportional to the mass of each body and inversely proportional to the square of the distance separating the bodies.

Heat. A quantitative measure of energy. The addition of heat to a body of matter causes an increase in internal energy. Removal of heat from a body causes the initial energy to decrease. The kilocalorie (kcal) unit of measurement in the metric system represents the amount of heat required to change the temperature of one kilogram of water by one degree Celsius. The British unit of heat energy is the (Btu) or British thermal unit.

Heat Engine. A device that converts internal (heat) energy into mechanical energy and whose behavior is governed by the laws of thermodynamics.

Heat of Combustion. The amount of heat released by the complete burning of a specific quantity of matter.

Heat of Fusion. The amount of heat needed to melt a definite mass of matter without changing its temperature.

Heat of Vaporization. The amount of heat needed to vaporize a definite mass of a liquid without changing its temperature.

Hooke's Law. The amount of stretch of certain materials (within a definite range) is proportional to the change in the amount of applied force.

Horsepower. A unit of measurement of power equal to 550 foot-pounds per second or 33 000 foot-pounds per minute, or 746 W.

Humidity. A condition indicating the presence of water vapor in a given space or area.

Impulse of a Force. The product of the force and the time during which it acts. The change in the momentum of a body that is free to move, due to a force acting on it, is equal to the impulse generated by the force.

Inertia. The tendency of a body to resist change in state, whether in motion or at rest.

International System of Units (SI). SI metrics, as used in this text, is an up-to-date international system of weights and measures. SI is a coherent system. There are seven base units with established names, symbols, and precise definitions.

Joule (J). An SI metrics unit of work or energy. One joule (J) is equal to 10^7 ergs, or approximately 0.739 0 gram calories, or 0.737 5 foot-pounds. $1 J = 1 kg \cdot m^2/s^2$.

Kelvin Degree (°K). The unit of temperature measurement in SI metrics. A degree on the Kelvin absolute scale is equal to the °C.

Kilo (k). An SI metrics prefix denoting one thousand times (10^3).

Kilocalorie (kcal). The unit of heat in the metric system. 1 kcal = 4185 J.

Kilogram (kg). The base unit of mass in SI metrics.

Krypton$_{86}$. An inert, colorless, gaseous element that provides the foundation for defining the meter in SI metrics.

Lines of Force. Denote the direction a body would move if released at a specific point. The concentration of the lines of force is proportional to the magnitude of the force.

Liter (L). A metric system unit of volume or capacity. One liter equals one cubic decimeter (dm^3) or 1000 grams.

Manometer. A common instrument for measuring fluid pressure; has a U-shape.

Mass (m). A quantitative measure of the inertia of a body at rest. The greater the resistance of a body to being set in motion, the greater is its mass. The kilogram (kg) is the metric unit of mass; the slug is the British system unit of mass.

Mechanical Advantage. The number of times the resistance force is greater than the effort force.

Mega (M). An SI metrics prefix which denotes one million times (10^6).

Meter (m). A base unit of length in SI metrics. One meter equals the wavelength of light which is 1 650 763.73 wavelengths of the colored line produced by krypton$_{86}$.

Metrication. A widely accepted term which applies to any process or program of conversion to the International System of Units, identified in this text as SI metrics.

Metric System. A designation for the French and European metric systems using the meter as the basic unit of measure.

Mole (mol). A base unit in SI metrics representing the amount of substance of a system.

Moments (Principle). The sum of the clockwise and counterclockwise moments is equal when a body is balanced (in equilibrium).

Momentum Arm of a Force. The perpendicular distance from a pivot to the line of action of the force.

Momentum, Linear. The product of the mass and velocity of a body.

Neutron. An uncharged particle found in the nucleus of an atom.

Newton (N). A derived unit of force in SI metrics. $1 \text{ N} = 1 \text{ kg} \cdot \text{m/s}^2$.

Nucleus. The core of every atom consisting of one or more neutrons and protons.

Parallelogram of Forces. A mathematical method of representing the direction and quantity of the forces acting on a point in a parallelogram.

Pascal (Pa). A derived unit of pressure in SI metrics. One pascal is equal to one newton per square meter. $1 \text{ Pa} = 1 \text{ N/m}^2$.

Pascal's Law. A law stating that the pressure on a confined fluid (at rest) is transmitted equally in all directions.

Period. The time required for a body undergoing simple harmonic motion to make one complete oscillation, or for one complete wave to pass a specific point.

Permeability. A measure of the magnetic properties of a medium.

Plasma. A gas composed of electrically charged particles whose behavior depends largely upon electromagnetic forces.

Potential Energy. A condition or position of energy to do work.

Pound, Avoirdupois. A unit of weight in the customary system. One pound avoirdupois equals 16 avoirdupois ounces, 0.453 592 37 kg or 7000 g.

Power. The rate at which work is done. The watt (W) is the unit of power in the metric system and ft-lb/s is the unit of power in the British system.

$1 \text{ W} = 1 \text{ J/s}$

$1 \text{ horsepower} = 746 \text{ W} = 550 \text{ ft-lb/s}$

Quart. A United States liquid unit equivalent to 57.75 cubic inches.

Radian. A unit of angular measure equal to $57.30°$. There are 2π radians in a full circle.

Resonance. A condition of response where the impressed frequency of one body is equal to the natural frequency of the other body which causes it to move.

Resultant. A single force having the same effect as two or more forces acting together at a point.

SI. A standard abbreviation for the International System of Units of measurement.

Significant Digit (Places). The decimal place or digit that is necessary to accurately define a quantity or value.

Slug. The unit of mass in the British system. One slug weighs 32 pounds.

Specific Gravity. A numerical value indicating the ratio of the weight of an object and an equal volume of water.

Specific Heat. The heat required to raise one pound of a specific material one degree Fahrenheit, or one gram one degree Celsius.

Surface Tension. A force exerted by the surface of a liquid which tends to pull it together.

Thermal Conductivity. The measure of the ability of a material to conduct heat.

Torque. The product of the magnitude of a force about a specific axis and the perpendicular distance from the line of action of the force to the axis.

Turboshaft Engine. A gas turbine consisting of a gas generator section, a reduction gear box to produce rotary shaft motion, and turbine wheels preceding the exhaust section to turn the drive shaft.

Vector. A line on a vector diagram to represent both the direction and quantity of a force.

Velocity (v). A quantitative measure (vector quantity) of the speed and direction of a moving body.

Venturi Action (Bernoulli's Principle). The velocity of a fluid in motion in a constricted tube is greatest where the pressure is the least, and conversely.

Watt (W). A unit of power. One watt equals one joule per second, or 1/746 horsepower.

Work. Work is a measure of the amount of change produced by a force acting upon an object. The amount of work is equal to the magnitude of a force multiplied by the distance through which it acts. The joule (J) is the unit of work in the metric system; the foot-pound (ft-lb) is the unit in the British system.

Index

Acceleration, 84-85
 and force, 86
 measurement, 48
Adhesion
 as cause of friction, 144
 definition, 10, 18
 industrial applications, 19
Air, weight of, 24-25, 163
Air pump, 166
Altimeter, 163
American National Standards Institute, 41
Ammonia gas, 25
Ampere, definition, 45
Amplitude, definition, 190
Angles, 46-47
ANSI. See American National Standards Institute
Archimedes' principle, 156
Area, 34-35, 47, 49, 52
Aristotle, experiments of, 82
Astronomy, definition, 3
Atmospheric pressure. See Pressure, atmospheric

Balance, dynamic and static, 72-74
Barometer, 24, 163-64
Base ten scientific notation system, 41, 42-43
Bath system of lubrication, 148
Bearings, and friction, 146-47
Bernoulli's principle, 180
 and fluids, 157
Biology, definition, 3
Block and tackle. See Pulley, mechanical advantage
Botany, definition, 3
Boyle, Robert, 164-65
Boyle's law, 165, 174
 industrial applications, 168
Buoyancy, 20, 156

Cams, 89, 106
Candela, definition, 46
Capillary action, 18-19
Carbon dioxide, compressibility of, 25
Center of gravity, 73

Centrifugal force, 87-88
Centripetal force, 87-88
Chemistry, definition, 3
Circuits, fluid, 176-78
Coefficient of friction, 145-46
Cohesion, 10, 18, 144
 industrial applications, 19
Component. See Force, component
Composition of forces method. See Parallelogram method of vector representation
Compressibility of gases, 23-24
Compression and force, 65
Computer, analog, 90
Conductivity, 15
Conversion, 50-54
Cubes, volume of, 35-36

Deceleration, uniform, 85
Deformation due to friction, 145
Density, 9-10, 25
Displacement, as vector quantity, 75
Displacement method of measuring volume, 36-37
Distance, mechanical advantage of, 104, 106
Ductility, 15
Dynamometer, 142

Efficiency, 65, 68
Effort, 95
Elasticity, definition, 15
Electricity, 48
Emulsions, industrial applications, 19
Energy
 of fluid systems, 178
 transfer, 184
Equilibrant, 77-78
Equilibrium, 72, 73-74

Flow, 174, 176, 179
Fluid circuits, 177
Fluid power, 174-75, 179-80
Fluid power systems, and expansion compensation, 175
Fluid pressure
 application to machines, 155-56

 engineering applications, 155
 and force, 153-54
 industrial applications, 168
 and size or shape of container, 154-55
Fluid resistance, result of, 176
Fluid systems, 177-79
Fluidic control devices, 180
Fluidics, application of, 180
Fluids, 154, 156, 176
 applications of, 167-68
Force
 angular, 76-77
 centrifugal and centripetal, 87-88
 compared with pressure, 154
 component, 78
 definition, 64
 effects of, 74
 effort and resistance, 65-66
 kinds of, 64-65
 magnitude of, 78
 measurement, 47
 mechanical advantage of, 66-67, 98-99, 103-4, 106, 111, 119-21
 moment, 72, 97
 parallel, 76
 point of application, 78
 resultant, 74-75
 torque, 72
 transmission of, 175-76
 as vector quantity, 75
Freely falling objects, 82-83
Frequency
 measurement, 47
 of waves, 189-90
Friction
 advantages and disadvantages, 143-44
 causes, 144
 coefficient of, 145-46
 deformation due to, 145
 internal, 176
 overcoming, 146-47
 types, 144-45
Fulcrum, 95, 110
Fusibility, definition, 15

Gage, venturi, 157

219

Galileo, experiments of, 82-83
Gases
  change to liquid or solid state, 25
  evaporation of liquids to form, 25
  properties of, 23-25
  result of pressure on, 174
Gear train
  compound, 129
  practical applications of, 130
  simple, 126, 128
Gears
  driver and driven, 127
  effect on force, 129
  in power transmission, 140
  rotation and, 126-27
  and speed, 127-28
Gravity, 83. See also Center of gravity; Specific gravity
Gravity pressure, 154

Hardness, 14
Helix angle of screw threads, 120-21
Hertz (Hz), 48
Horsepower, compared with work, 67-68
Hydraulic press, 155-56
Hydraulic systems, 174-76
  uses of, 179-80
Hydrometer, 156

Impenetrability, 10
Inclined plane, 103-5
Inertia, 10, 86, 87
International metric units, using, 41-59
International Standards Organization, 41
International System of Units
  adoption of, 41
  advantages, 43
  base ten scientific notation system, 41
  base units, 44-46
  basic measurement units in, 47
  common metric units, symbols and formulas, 49
  conversion factors, 50-54
  derived units, 47-49
Invariant quantity in nature, 32
ISO. See International Standards Organization

Law of Moments, 97

Length, measurement of, 33
Levers
  characteristics, 95
  classification of, 96-97
  industrial applications, 99
  mechanical advantage of, 98-99
  types, 96-97
Linear measure. See Length, measurement of
Liquids
  with greater adhesion than cohesion, 18-19
  with greater cohesion than adhesion, 19
  properties, 18-19
  results of pressure on, 174
Lubrication, 147-48
Luminous intensity, derived units of, 48

Machinability, definition, 15
Machines, 95, 141-42
  compound, 134-35
  ideal, 156
  simple, 133
Magnitude. See Force, magnitude of
Malleability, 14
Mass
  definition, 9
  and momentum, 86
Matter
  physical properties, 8-11
  states of, 8, 14
  structure of, 8
Measuring devices, 118
Mechanical advantage
  of compound gear train, 129-30
  of compound machines, 134-35
  of force, 66-67
  of hydraulic press, 155-56
  of inclined plane, 103-4
  of levers, 98
  of pulley, 133-34
  of screw threads, 119-21
  of speed, 66-67
  of wedge, 106-7
  of wheel and axle, 111-12
  of worm and worm wheel, 135
Mechanical movements, 88-90
Mechanical power, 139
Mercury, cohesion and adhesion of, 19
Metallurgy, definition, 3

Metrication, 41-59
Micrometer, 33-34
Molecular attraction, 10
Molecular motion, 23
Molecules, 8, 10, 206
Moment, 72-73
Moment of force, 110-11
Momentum, factors controlling, 86-87
Motion
  common concepts of, 83-84
  fluid, 178
  laws of, 86-87
  positive rotary, 126
  rectilinear, 89, 117
  rotary, 89, 117-18
  transmitting, 89
Movement, mechanical advantage of, 98-99

Newton Sir Isaac
  law of gravitation, 83
  laws of motion, 86-87
Nylon, development of, 1

Oil, miscible, 19
Opposition, 176. See also Resistance; Reluctance; Friction; Fluid resistance
Parallelogram method of vector representation, 77
Pascal's Law, 154-55, 156-57
Physical change, definition, 8
Pneumatic systems, 174, 176, 179-80
Point of application of forces, 78
Porosity, 10
Power
  continuous transmission, 140
  coupling devices, 140
  definition, 67, 139
  intermittent, 140
  measurement, 47, 141-43
  mechanical transmission of, 139-40
  sources, 139
  standard unit of measure, 67
Power systems, principle governing, 176
Power transmission, through fluid system, 175
Pressure. See also Pressure, atmospheric; Pressure, fluid

and compressibility of gases, 23-24
  measurement, 47
Pressure, atmospheric, 24
  barometers and, 163
  basic principles of, 164-67
  Boyle's Law, 165-66
  and fluid movement, 167-68
  industrial applications, 168-69
  measurement of, 163-64
Pressure, fluid, 154-55
  Pascal's law, 154
Pressure system, of lubrication, 148
Problem solving, systematic
    approach to, 4-5
Prony brake, 142-43
Pulley
  mechanical advantage, 133-34
  in power transmission, 139-40
  types, 133
Pulses, 186-87
  superposition of, 187-88
Pumps
  fluid, 168
  force, 167
  lift, 167

Radian. See Angles
Rarefaction pulse, 185
Reluctance, in magnetic circuits, 176
Reservoir, for fluid systems, 178
Resistance
  to flow, 176
  of fluid systems, 177
  and levers, 95
  parallel circuit and, 178
  series circuit and, 177
  of series-parallel system, 178
  thermal, 179
Resultant, definition, 74
Rules, steel, 33

Scalar, 74-75
Science, importance of, 1-3
Scientific laws, 4
Scientific method, 4
Screw threads
  applications, 118-19, 122
  characteristics of, 121
  as fastening devices, 119

force transmitted with, 119-21
motion transmitted with, 117-18
terms, 117
uses of, 117-21
Second, definition, 45
Sheave, 133
SI. See International System of
    Units
SI metrics. See International System
    of Units
Siphon, and atmospheric pressure,
    166-67
Slope of inclined plane, 103
Solids
  molecular arrangement in, 14
  properties, 14-15
Sound, speed of, 89
Specific gravity, 20, 156
Specific weight of gases, 25
Speed, 84-86
  change in, using wheel and axle,
    111-12
  control of, 112-13
  mechanical advantage of, 66-67,
    98-99, 129-30
Stability. See Equilibrium
Steady-state, in fluid systems, 179
Surface tension, 18
Systeme International d'Unites (SI).
    See International System of
    Units

Tapes, steel, 33
Temperature
  effect on gas pressure, 24
Tenacity, definition, 15
Tension, and force, 64
Time constant, of fluid systems, 179
Torque. See also Force
  formula for, 66
  of lever, 99
Torricelli, Evangelista, 163
Torricelli barometer, 163
Toughness, definition, 14

U.S. Metric Board, 41

Vector
  definition, 74

graphic representation, 77
quantities, 75
Velocity
  measurement, 48
  and momentum, 86-87
  as vector quantity, 75
Viscosity, 19-20
  of fluids, 176
Volatile liquids, storage of, 25
Volume
  definition, 9

Water, 19
Wave motion, 184
  pulse of, 186-87
Wave train, 189. See also Pulses
Waveform, 189
Wavelength, 189-90
Waves
  amplitude, 190-91
  electromagnetic, 184
  frequency of, 189-90
  interference of, 188-89
  mechanical, 184
  period of, 189-90
  periodic, 189-90
  in phase, 189
  transverse, 184-85
  velocity of, 185, 189-90
Wedge
  compared with inclined plane, 105
  mechanical advantage of, 106
  uses, 106
Weight, 9, 37
Wetting action, 18. See also
    Adhesion of liquids
Wheel and axle, 110-11
  changing speeds using, 111-13
  gears as application of, 129
  industrial applications, 112-13
Work, 64, 67
Worm and worm wheel, 134-35

Yardstick, 33

Zoology, definition, 3